J. Ellard Gore

An astronomical glossary, or dictionary of terms used in astronomy

With tables of data and lists of remarkable and nteresting celestial objects

J. Ellard Gore

An astronomical glossary, or dictionary of terms used in astronomy
With tables of data and lists of remarkable and nteresting celestial objects

ISBN/EAN: 9783742827234

Manufactured in Europe, USA, Canada, Australia, Japa

Cover: Foto ©ninafisch / pixelio.de

Manufactured and distributed by brebook publishing software
(www.brebook.com)

J. Ellard Gore

An astronomical glossary, or dictionary of terms used in astronomy

AN

ASTRONOMICAL GLOSSARY

OR

Dictionary of Terms Used in Astronomy

*WITH TABLES OF DATA AND LISTS OF
REMARKABLE AND INTERESTING
CELESTIAL OBJECTS*

BY

J. E. GORE, F.R.A.S., M.R.I.A., ETC.

*Fellow of the Imperial Institute, Honorary Member of the Liverpool Astronomical
Society, Corresponding Member of the Astronomical and Physical Society
of Toronto. Author of " Planetary and Stellar Studies," " The
Visible Universe," " The Scenery of the Heavens," etc.*

Capio Lumen

LONDON

CROSBY LOCKWOOD AND SON

7, STATIONERS' HALL COURT, LUDGATE HILL

1893

PREFACE.

The following Glossary contains an explanation of all the terms and names generally used in books on Astronomy, and it is hoped that it will be found useful as a work of reference both to the beginner and the advanced student.

Tables are added containing the latest values of Astronomical Constants, details of the Planets and Satellites of the Solar System, and lists of Remarkable Red, Variable, and Binary Stars.

J. E. G.

October, 1893.

CONTENTS.

AN ASTRONOMICAL GLOSSARY.

A.

Aberration of Light. An apparent displacement in the position of the stars due to the effect of the earth's motion in its orbit round the sun combined with the progressive motion of light. The result is that "a star is displaced by aberration along a great circle joining its true place to the point on the celestial sphere towards which the earth is moving" (Barlow and Bryan's *Mathematical Astronomy*, p. 298). The amount of aberration is a maximum for stars lying in a direction at right angles to that of the earth's motion. This is known as the "constant of aberration," and its value in seconds of arc is 206,265 multiplied by the velocity of the earth and divided by the velocity of light, or about 20·5″. The motion of the earth on its axis also produces a small aberration called the Diurnal Aberration, but the coefficient of this is very small—only 0·32″—and almost imperceptible in observations. For a star on the celestial equator, viewed from the earth's equator, the time of transit would be retarded by diurnal aberration by only $\frac{1}{30}$th of a second, which could hardly be observed.

Absorption of Light. A supposed diminution in the brightness of very distant stars by absorption of their light in the luminiferous ether of space. It is also termed the "extinction of light." That such an

1

absorption of light really takes place in the ether has not, however, been well established.

Acceleration. Secular of moon's mean motion. A slow increase in the velocity of the moon's mean orbital motion round the earth due to the change in the eccentricity of the earth's orbit round the sun.

Achernar. A name applied to the star α Eridani. Derived from the Arabic *âchir al-nahr*, "the end of the river" (Eridanus).

Achromatic. A refracting telescope in which the lenses are so constructed that an image of an object practically free from colour is formed.

Acolyte (an attendant). A term sometimes applied to a faint star seen in the same field of view with a much brighter one.

Acronical. When a celestial body rises or sets with the sun it is sometimes said to rise acronically.

Adara. A name sometimes applied to ε Canis Majoris. Derived from the Arabic *al-adzârî*, "the virgins," a term applied by the old Arabian astronomers to the stars o², δ, ε, and η Canis Majoris.

Aërolite. A term applied to a kind of meteoric stones which occasionally fall from the sky, and which are composed almost entirely of stone, with little or no iron.

Æther. See ETHER.

Aish. An ancient name for the Great Bear or Plough.

Albedo of a Planet. The proportion of the sunlight reflected from a planet's surface compared with the total amount received from the sun.

Albirco. A name applied to the star β Cygni.

Alchiba. A name sometimes applied to the star α Corvi.

Alcor. A small star closely *following* the star ζ

Ursæ Majoris (Mizar). It is otherwise known as *g* or 80 Ursæ Majoris.

Alcyone. A name applied to the star η Tauri, the brightest star in the Pleiades.

Aldebaran. A name applied to the first magnitude star α Tauri. Derived from the Arabic *al-dabarân*, " the follower," because it follows the Pleiades.

Alderamin. A name applied to the star Cephei. Derived from the Arabic *al-dzirâ al-jumîn*, "the right arm " (of Cepheus).

Aldhibaïn. A name applied by the Arabian astronomers to the stars η and ζ Draconis. The word means " the two jackals."

Alfeta. A name given in the Almagest to the star α Coronæ Borealis.

Algeiba. A name sometimes applied to the star γ Leonis.

Algenib. A name sometimes applied to the star γ Pegasi. Probably a corruption of the Arabic *djanâh al-farras*, " the wing of the horse."

Algol. From the Arabic *ras al-gûl*, " the head of Algol " (Medusa). The famous variable star β Persii. The period of variation from minimum to minimum was, in 1696, about 2 days 20 hrs. 48 mins. 59 secs., but has now diminished to 2 days 20 hrs. 48 mins. 51 secs. The star remains constant in light for the greater portion of its period, and the whole of the light fluctuations take place in a period of about 10 hours. The variation of light is from 2·3 to 3·5 magnitude.

Algorab. A name sometimes applied to the star α Corvi.

Algores. A name sometimes applied to the star δ Corvi.

Alhena. A name sometimes applied to the star γ Geminorum. From the Arabic *al-hanat*.

Alioth. A name sometimes applied to the star ε Ursæ Majoris.

Alkaid. A name sometimes applied to the star η Ursæ Majoris.

Alkalurops. A name sometimes applied to the star μ Boötis.

Alkes. A name sometimes applied to the star α Crateris.

Almack. A name sometimes applied to the star γ Andromedæ.

Alnilam. A name sometimes applied to the star ε Orionis.

Alnitak. A name sometimes applied to the star ζ Orionis.

Alphard. A name sometimes applied to the star α Hydræ. From the Arabic *al-fard*, "the solitary one," because there is no other bright star near it.

Alphecca. A name applied to the star α Coronæ Borealis, "the gem of the coronet." From the Arabic *al-munîr min al-fakka*, "the brilliant of the crown."

Alpherat. A name applied to the star α Andromedæ.

Alphirk. A name sometimes applied to β Cephei.

Alshain. A name sometimes applied to β Aquilæ.

Altair. A name applied to the bright star α Aquilæ. From the Arabic *al-tâir*.

Altitude. The angular elevation of a star above the horizon, measured on a great circle passing through the star and zenith. The measured angle must be corrected for the effect of refraction, which tends to apparently raise the star above its true position.

Altitude and Azimuth Instrument. A telescope

mounted so as to be movable about a horizontal and also a vertical axis. It is also called an *altazimuth*. Small telescopes are usually mounted in this way. Tho theodolite, used in surveying, is an altazimuth.

Aludra. A name sometimes applied to the star η Canis Majoris.

Alwaid. A name sometimes applied to the star β Draconis. It is derived from the Arabic *al-awâidz*, "the old camels," a term given by the Arabians to tho stars ν, β, ξ, and γ Draconis, which form, with ι Herculis, the well-known cross, marking the head of the Dragon and the left foot of Hercules.

Amplitude. The angular distance of a celestial body when rising or setting from the east or west points of the horizon. The amplitude is measured on the horizon.

Andromeda (the Chained Lady). One of the northern constellations.

Andromeda Nebula. The great nebula in Andromeda known to astronomers as 31 Messier. It lies closely preceding the 4½ magnitude star ν Andromedæ. It is visible to the naked eye on a clear, moonless night, and is a conspicuous object with small telescopes, and even with a good binocular field-glass. It seems to have been familiar to the ancients, as it is mentioned by tho Persian astronomer Al-Sufi, who wrote a description of tho heavens in the tenth century.

Andromedes. A meteor shower, visible about November 27th in each year. It appears to radiate from a point near γ Andromedæ (25° + 43°). Tho meteors are very slow and trained.

Angle. The inclination of one straight line to another.

Angle of Eccentricity. In an ellipse, the angle

between the minor axis and a line drawn from the extremity of the minor axis to the focus of the ellipse. The size of this angle is the eccentricity of the ellipse.

Angle of Position. The position of the line joining the components of a double star, with reference to the circle of declination passing through the principal star of the pair. The zero is at the north point, and the angles are measured from 0° to 360°, from the north point round by east, south, and west. For example, if the position angle is 90°, the companion is due east of the primary stem; if 180°, it is exactly south of it ; and if 270°, it is due west.

Angle of Situation. The angle between the circles of declination and of latitude passing through a given star.

Angular Velocity. The rate at which the angle described by the radius vector of a moving body changes. See RADIUS VECTOR.

Annual Equation. An inequality in the moon's motion, due to the varying distance of the earth from the sun.

Annular Eclipse. A solar eclipse in which the sun is only partly covered by the moon, a ring or annulus of sunlight being left uncovered round the moon's disc.

Annular Nebulæ. Nebulæ of a ring-shaped form. They are among the rarest of celestial objects. The most remarkable object of this class is that situated between the stars β and γ Lyræ, and known to astronomers as 57 Messier.

Annular Variation. The correction to be applied per annum to the right ascension and declination of a star, due to the effects of precession and the star's proper motion.

Anomalistic Month. See MONTH, ANOMALISTIC.

Anomalistic Year. Tho time which elapses between two successive passages of the Sun (in its apparent revolution among the stars) through the perigee of the earth's orbit. The length of the anomalistic year is 365 days 6 hrs. 13 mins. 49 secs.

Anomaly of a Planet. The angle between the place of a planet and the major axis of its orbit is called its anomaly. This angle is measured in three ways, which are known as the eccentric, mean, and true anomalies.

Anomaly, Eccentric. An auxiliary angle used in the calculation of the orbits of planets and binary stars. If a circle be described on the major axis of the elliptic orbit, and a perpendicular be drawn to the major axis through the true place of the moving body; then, if the point where this perpendicular meets the auxiliary circle be joined with the centre, the eccentric anomaly is the angle between this line and the major axis.

Anomaly, Mean. The angle between the perihelion and the mean place of a planet, comet, or the component of a binary star. The " mean place " at any given time is tho place which the body would occupy if it revolved round its primary in a circular orbit with a uniform velocity, and with a period of the same length as that in the real orbit.

Anomaly, True. The angle between the perihelion of a planet and its true place in the orbits of planets and comets, or the angle between the periastron and the true place of the companion in the real orbit of a binary star, is termed the *true anomaly*.

Ansæ (Handles). A term applied to the portions of Saturn's rings which appear to project on each side (due to perspective) of the planet's globe. To the old astronomers, with their imperfect telescopes, these appeared like handles to the ball : hence the name.

Ant-apex. The point in the celestial sphere *from* which the sun is moving in space.

Antarctic Circle. The circle on the earth's surface ni the southern hemisphere, which lies $23\frac{1}{2}°$ from the south pole, or of which the latitude is $66\frac{1}{2}°$ south.

Antares. A name applied to the red star a Scorpii.

Antlia. One of the southern constellations. It lies south of Hydra and north of Vela (Argo).

Aperture. The diameter of the object-glass of a refracting telescope, or of the mirror in a reflecting telescope.

Apex. A term usually applied to denote the point in the sky towards which the sun is moving in space. This point is called the "Solar Apex." Various determinations of its position have been made; but most of the points found lie in Hercules and Lyra. The term "apex" is also sometimes applied to the point on the ecliptic towards which the earth's orbital motion round the sun is directed at any instant. This point lies 90° from the sun towards the west, and is called "the apex of the earth's way."

Aphelion. The point in the orbit of a planet or comet which is most remote from the sun. This point lies at the extremity of the major axis of the ellipse, and nearest the focus which is sometimes called "the empty focus."

Aplanatic. A term applied to a telescope or other optical instrument in which the chromatic and spherical aberrations have been satisfactorily corrected by a combination of suitable lenses. The construction of an *absolutely* aplanatic instrument is probably impossible.

Apoastron or **Aphastron.** The point in the *real* orbit of a binary star at which the components are farthest from each other. This point does not always coincide with the point of maximum distance as measured in the *apparent* orbit. The apoastron point may be found by drawing a line from the primary star to the centre of the apparent ellipse, and producing it to meet the ellipse. The opposite intersection of this line with the apparent ellipse is the periastron point.

Apogee. The point in the moon's orbit which is farthest from the earth.

Apparent Ellipse. The ellipse described by one component of a double star round the other *as seen from the earth.* The apparent ellipse is the orthogonal projection of the real ellipse on the background of the sky.

Apparent Motion. The motion of a celestial body as seen from the earth. The term is sometimes applied to the apparent diurnal motion of the heavenly bodies, due to the earth's rotation on its axis, and sometimes to the motions of the sun, moon, and planets among the fixed stars on the celestial sphere.

Apparent Sun. A term applied to the sun itself, or "true sun," to distinguish it from the imaginary, or "mean sun."

Apparition, Circle of Perpetual. See CIRCLE OF PERPETUAL APPARITION.

Appulse. The apparently close approach of two celestial bodies.

Apse, or **Apsis.** A term applied to the "perigee" and "apogee" of the orbit of the earth and moon, or the "perihelion" and "aphelion" of a planet's orbit.

Apsides, Line of. The line joining the "perigee" and "apogee" of the earth's orbit round the sun, of the moon's orbit round the earth, or the "perihelion" and "aphelion" of a planet's orbit.

Apus (the Bird of Paradise). A southern constellation. It lies between Triangulum Australis and the southern celestial pole.

Aquarids. Meteor showers visible about May 1st and July 27th to 29th in each year. They seem to radiate from points in the constellation Aquarius (326°−2° and 341°−2°). In both showers the meteors have long paths, but those in May are swift and those in July are slow.

Aquarius (the Water Bearer). One of the zodiacal constellations.

Aquila (the Eagle). One of the constellations. The celestial equator passes through it. Its brightest star is Altair (Aquilæ).

Ara (the Altar). One of the southern constellations.

Arc. A portion of any curve.

Arc, Diurnal. The portion of a circle parallel to the equator which is described by a celestial body between its rising and its setting.

Arc of Progression. The arc in the sky described by a planet when in direct motion—that is, from west to east, or in the order of the signs of the zodiac.

Arc of Retrogradation. The arc described by a planet when apparently retrograding—that is, moving from east to west, or contrary to its real motion in space. This apparent retrograde motion is due to the earth's motion round the sun combined with the motion of the planet.

Arctic Circle. The circle on the earth's surface in the northern hemisphere which lies $23\frac{1}{2}°$ from the north pole, or of which the latitude is $66\frac{1}{2}°$ north.

Arcturus. A name applied to the bright star *α* Boötis.

Areal Velocity. The area of the sector traced out by the "radius vector" of a moving body in the unit of time. This area is equal to half the linear velocity multiplied by the perpendicular from the centre of force, or the tangent at the given point.

Areas, Kepler's Law of. When one body revolves round another as a centre of force, the radius vector, or line joining the two bodies, traces out equal areas in equal times. This law applies to the motion of the earth and planets round the sun, the satellites round the planets, the components of binary stars round each other : in fact, it holds true in the case of any body moving round a centre of force under *any* law of force.

Argo (the Ship Argo). A large constellation in the southern hemisphere. It is usually subdivided into four divisions : Puppis, Malus, Vela, and Carina. Its brightest star is Canopus, which ranks only second to Sirius in brilliancy.

Arided. A name sometimes applied to the star *α* Cygni.

Ariel. The inner satellite of Uranus, or that nearest to the planet. Its mean distance from the planet's centre is about 127,000 miles, and its period of revolution 2 days 12 hrs. 29 mins. It can be well seen only with a large telescope, and its diameter is uncertain. Ariel was discovered by Lassell on Sept. 14th, 1847.

Aries. One of the zodiacal constellations. When the signs of the zodiac were established, the *vernal* equinox was situated at the beginning of this constellation, but owing to the precession of the equinoxes, the point of intersection of the ecliptic and equator has now retrograded into Pisces. The point is, however, still termed *the first point of Aries.*

Armillary Sphere. An ancient instrument constructed with metallic circles representing the astronomical circles of the celestial sphere.

Arnab. A name sometimes applied to the star α Leporis.

Artificial Horizon. A box containing mercury, which forms an horizon when observing altitudes of the celestial bodies. With an artificial horizon there is no *dip* to be taken into account (see DIP OF HORIZON), and the observed angle is double of the real angle of elevation or altitude.

Ascension, Right. The angular distance of a celestial body, measured from the " First Point of Aries " eastwards on the equator. This, combined with the declination, which is measured north and south from the equator on a great circle passing through the celestial poles and the body, determines the position of the object on the celestial sphere.

Asell Australis. A name applied by the ancient Romans to the star δ Cancri.

Asterism. A constellation or group of stars.

Asteroids, or Minor Planets, which see.

Asterope. One of the stars in the Pleiades.

Astræa. One of the minor planets which revolve round the sun in orbits lying between those of Mars and Jupiter. It was discovered by Hencke on Dec. 8th, 1845. It revolves round the sun in about 4·14 years, at a mean distance of 2·578 times the earth's mean distance from the sun. Even when favourably situated, its magnitude does not exceed the ninth, and its real diameter does not probably exceed sixty miles.

Astral. Having relation to the stars.

Astrolabe. An instrument invented by Hippardens to show the circles of the celestial sphere.

Astrology. The so-called science of predicting future events by the positions and aspects of the sun, moon, and planets.

Astrometer. An instrument for measuring the relative brightness of the stars. The term *photometer* is now generally used.

Astrometry. The measurement of the relative brightness of the stars. Now called *photometry*.

Astronomical Clock. A clock used in astronomical observations and regulated to show sidereal time. It therefore gains nearly four minutes a day on an ordinary clock, or 24 hours in the year. The dial is divided from 0 hrs. to 24 hrs., and the hands should point to 0 hr. 0 min. 0 sec. when the " First Point of Aries " transits the meridian.

Astronomy. The science which deals with the heavenly bodies, their distances, magnitudes, and motions, and the laws which govern them. It is derived from two Greek words—ἀστήρ, a star, and νόμος, a law—the law of the stars.

Atlas. One of the stars in the Pleiades. Otherwise known as 27 Tauri. The term is also applied to a set of star maps.

Atmosphere. The gaseous envelope surrounding the earth and some, at least, of the planets of the solar system. The earth's atmosphere consists of a mechanical mixture of oxygen and nitrogen gases, the proportions by volume being 79 parts nitrogen and 21 parts oxygen, with a very small quantity of carbonic acid gas. It probably extends to a height of 100 miles or more above the earth's surface, with a constantly diminishing density; but the pressure is equivalent to that of a homogeneous atmosphere of about $5\frac{1}{4}$ miles in height and of a density equal to that at the earth's surface.

Attraction of a Sphere. The attraction of a sphere on a body outside it is the same as if the whole mass were collected at the centre of the sphere.

Augmentation of Moon's Apparent Diameter. The increase in the moon's apparent diameter due to an observer on the surface of the earth being nearer to the moon than the earth's centre, to which mathematical calculations are referred.

Auriga (the Waggoner or Charioteer). One of the northern constellations. Its principal star, Capella, is one of the brightest stars in the northern hemisphere, and is about twice as bright as an average star of the first magnitude.

Aurora Borealis, or Northern Lights. A luminous phenomenon visible in the atmosphere in the arctic regions, and occasionally, to some extent, in more southern latitudes. A similar phenomenon occurs near the south pole. Aurorae are supposed to be caused by electrical discharges in the upper regions of the earth's atmosphere.

Australis, Asad. A name sometimes applied to the star ϵ Leonis.

Autumnal Equinox. The equinox at which the sun passes from the north to the south side of the equator. This takes place on September 23rd. See EQUINOXES.

Axis of an Orbit. This term is applied to the major axis of the ellipse in elliptical orbits. It is also the line of apsides.

Axis of a Planet. An imaginary line through the planet, round which it rotates.

Axis of Figure. The solids formed by the revolution of a given surface round a fixed line. This line is called the axis of figure. Thus, an oblate spheroid may be supposed to be generated by the rotation of an ellipse round its minor axis. In this case the minor axis is the axis of figure.

Axis of Rotation. The axis round which a body rotates. In the case of the earth recent researches seem to show that the axis of rotation does not coincide *exactly* with the axis of figure. The difference is, however, very small.

Azelfafage. A name sometimes applied to the star π' Cygni.

Azha. A name sometimes applied to the star η Eridani. The Arabic word is *Udh-ha*.

Azimech. The star Spica (α Virginis) is sometimes called *Spica Azimech*.

Azimuth. The angle between the meridian and the great circle passing through the zenith and any given celestial body is called the *azimuth* of the body.

B.

Barlow Lens. A miniature achromatic object-glass with a negative focus, sometimes used in telescopes to increase the power of the eyepiece. It is placed between the object-glass and eyepiece, a few inches behind the eyepiece.

Base-line. A carefully measured line used in trigonometrical surveying, and also in the calculation of the distances of the heavenly bodies. In finding the sun's distance the base line is the earth's semi-diameter or radius, and in the determination of stellar distances the base line is the radius of the earth's orbit, or the mean distance of the earth from the sun.

Baten Kaitos. A name applied by the Arabian antronomers to the star ζ Ceti.

Beads, Baily's. A broken line of light seen on the sun's limb immediately before the totality of a solar eclipse. They are so called after the astronomer Francis Baily, who described them in 1836, but they were first seen by Halley during the total eclipse of 1715. A similar appearance has also been seen at the end of the total phase, and also in annular eclipses.

Beïd. A name applied by the Arabian astronomers to the star o Eridani. The word signifies an egg, and is supposed to have been given to the star on account of its white colour.

Bellatrix. A name applied to the star γ Orionis.

Belts. A term applied to the dark-coloured bands visible with a telescope on the discs of Jupiter and Saturn.

Benetnasch. A name sometimes applied to the star η Ursæ Majoris. From the Arabic *sarîr banâtnasch*.

Berthon's Dynamometer. An instrument for measuring the power of the eyepiece of an astronomical telescope, invented by the Rev. E. L. Berthon.

Bessel's Day Numbers. See DAY NUMBERS.

Bestiary. A name formerly applied to the zodiac.

Betelgeuse. A name applied to the red and vari-able star α Orionis. From the Arabic *ibit-al-djauzâ*.

Bifid. A term applied to comets' tails when they appear divided into two portions along their length.

Binary Stars. Double stars in which the com-ponents revolve round each other, or rather round their common centre of gravity. The number of known binary stars is now very considerable—probably not far short of a thousand. But, owing to the small arc of the orbit described by most of them since their discovery, it has been found possible to compute the orbit only in a limited number of cases. The orbits of about seventy binary stars have now been fairly well determined, some with considerable accuracy (one or more complete revolutions having been described). The periods of revolution vary in length from about $11\frac{1}{2}$ years to over 1600 years. (For List, see Appendix.)

Binocular. A form of telescope or large opera-glass having two tubes, used with both eyes at the same time.

Binuclear. A term applied to nebulæ which have two *nuclei* or condensations of light.

Bissextile. A term sometimes applied to Leap Year, in which a day is added to the month of February every four years.

Black Drop. An optical effect sometimes noticed in transits of Venus. Just after internal contact at ingress, and just before internal contact at egress, the planet has been seen in some transits apparently attached to the sun's limb by a dark ligament, probably the effect of irradiation and imperfect telescopes.

Bode's Law.* An empirical law connecting the distance of the planets from the sun. The law is as follows :—To each of the series of numbers 0, 3, 6, 12, 24, 48, 96, 192, 384 (in which each number—after the second—is double the preceding number) add 4, and we obtain the series 4, 7, 10, 16, 28, 52, 100, 196, 388, which represent approximately the distances of the planets from the sun, with the exception of Neptune, for which the distance indicated is considerably too large. The Earth's

---

* Bode's Law was really discovered by Titius.

distance being taken as 10, that of Neptune is about 300.

Bolides. A name applied to the large meteors; also known as fireballs.

Boötes (the Herdsman). One of the northern constellations. Its brightest star is Arcturus, one of the brightest stars in the heavens.

Borda's Principle of Repetition. A method of obtaining a more accurate measure of an angle by repeating the measure several times, and taking a mean of the measures. This is supposed to eliminate the errors due to imperfect graduation of the measuring circles, but in practice is not found very satisfactory, owing probably to imperfect clamping.

Box Sextant. A miniature form of sextant, chiefly used in surveying.

C.

Cælum (the Sculptor's Tool). One of the southern constellations.

Calendar (Gregorian). The omission of three leap years in every four hundred years was proposed by Pope Gregory XIII., and is called the *Gregorian correction.* According to this calendar, every year which is a multiple of 100 and is divisible by 400 is a leap year, and those not divisible by 400 are not leap years. Thus 1700, 1800, and 1900, are not leap years; but 2000 will be a leap year. The correction proposed by Pope Gregory leaves a small difference between the tropical year and the average civil year of about 1·23 day in 4000 years; but this may be safely neglected.

Calendar (Julian). The introduction of a leap year (a year with an additional day) every four years was due to Julius Cæsar, B.C. 44, and the calendar so constructed is called the Julian Calendar.

Calendar Month. The month used for the ordinary purposes of life. April, June, September and November have thirty days each, February has twenty-eight (and in leap year twenty-nine), "and all the rest have thirty-one."

Camelopardalis (the Giraffe). One of the northern constellations.

Cancer (the Crab). One of the zodiacal constellations.

Canes Venatici (the Hunting Dogs). One of the northern constellations.

Canis Major (the Great Dog). One of the southern constellations. Its principal star is Sirius, the brightest star in the heavens.

Canis Minor (the Little Dog). One of the constellations. Its principal star is Procyon, one of the brightest stars in the sky.

Canopus. A name applied to the bright southern star α Argûs. It ranks only second to Sirius in brilliancy. Derived from the Arabic word *Kânupus.*

Capella. A name applied to the bright star α Aurigæ.

Caph. A name sometimes applied to the star β Cassiopeiæ.

Capricornus (the Goat). One of the zodiacal constellations.

Cardinal Points. The cardinal points are north, south, east, and west. The north and south points are where the meridian meets the horizon. The east and west points are the intersections of the celestial equator with the horizon.

Carina (the Keel). A name applied to a part of the southern constellation Argo.

Cassegrainian Telescope. A form of reflecting telescope in which the smaller mirror is *convex*, and the reflected rays pass through a circular aperture in the large mirror.

Cassiopeia. One of the northern constellations. Popularly spoken of as " Cassiopeia's Chair."

Castor. A name applied to the star *a* Geminorum. It is a remarkable double and binary star.

Catoptrics. A division of the science of optics which deals with images formed by reflection from mirrors.

Cavendish Experiment. An experiment devised by Michell, and carried out by Cavendish in 1798, for the purpose of determining the density of the earth. The apparatus consists of two small equal balls placed at the extremities of a wooden rod, and suspended from the centre by a thin wire. The attraction of two heavy spheres placed alternately on opposite sides of the small balls produces a torsion in the suspending wire, the amount of which can be calculated. By observing the time of a small oscillation of the rod when acted on by gravity alone, and comparing this with the former result, the earth's density can be computed. See DENSITY OF EARTH.

Cebalrai. A name sometimes applied to the star *β* Ophiuchi.

Celæno. One of the stars in the Pleiades.

Celestial Equator. The great circle in which the plane of the terrestial equator meets the star sphere.

Celestial Globe. A globe on which the stars and constellations are depicted. In examining such a globe it should be remembered that the stars are drawn as supposed to be seen by an eye placed at the centre of the globe. The constellations are therefore reversed, and cannot be compared directly with the sky unless the inversion is mentally corrected.

Celestial Horizon. See HORIZON, CELESTIAL.

Celestial Latitude. The angular distance of a celestial body from the ecliptic, measured on a great circle at right angles to the ecliptic.

Celestial Longitude. The angular distance from the " First Point of Aries " to a " secondary " to the ecliptic passing through a given star. Celestial longitude is measured eastwards from the ecliptic.

Celestial Meridian. The great circle of the celestial sphere which passes through the zenith, nadir, and celestial poles.

Celestial Poles. The points in the celestial sphere towards which the earth's axis of rotation points. They are, in fact, the extremities of an imaginary axis round which the star sphere *apparently* rotates.

Celestial Sphere. The hollow sphere on the surface of which the heavenly bodies seem to be placed. The observer's eye is practically situated at the centre of the sphere, and consequently a complete hemisphere is always visible to the observer at any point on the earth's surface.

Centaurus (the Centaur). One of the southern constellations.

Centre of Ellipse. The middle point of the major axis of an ellipse, or the point where the major and minor axis intersect at right angles.

Centre of Figure. The centre of a regular solid, such as the sphere, ellipsoid, etc., is called the *centre of figure.*

Centre of Mass. The " centre of gravity " of a body is sometimes called the centre of mass. In a homogeneous sphere this will coincide with the centre of the sphere or the centre of figure ; but, if the sphere is not of the same density throughout, the centre of mass will not coincide with the centre of figure.

Centrifugal Force. If a body of mass, m, revolves in a circle of radius, r, with a velocity, v, the centrifugal force is $\dfrac{mv^2}{r}$, and acts outwards from the centre.

Centripetal Force. The force acting *towards* the centre which balances the centrifugal force.

Cepheus (the Monarch). One of the northern constellations.

Ceres. One of the minor planets revolving round the Sun in orbits lying between those of Mars and Jupiter. It was discovered by Piazzi at Palermo on January 1st, 1801 (the first day of the nineteenth century). It revolves round the Sun in a period of 4·60 years, at a mean distance of 2·767 times the Earth's mean distance from the Sun. The eccentricity of the orbit is small—only 0·076. Its opposition magnitude is about 7·7, and its real diameter perhaps about two hundred miles.

Cetus (the Whale). One of the constellations.

Chamæleon (the Chameleon). One of the southern constellations.

Chaph. A name sometimes applied to the star β Cassiopeiæ.

Cheliab. A name given in the Arabo-Latin Almagest to the constellation Perseus.

Chimah. An ancient name for the constellation Taurus.

Chinese Annals. Records of astronomical phenomena have been kept in China for many hundreds of years. These are generally referred to in books on astronomy as " the Chinese Annals of Ma-tuoan-lin."

Chronograph. An instrument devised for recording the times of star transits across the wires of a transit instrument. It was designed to supersede the old " eye and ear method." The instrument consists of a cylinder covered with paper, which is made to revolve steadily by clockwork and pushed forward by a screw on the axle. The record is made by means of a pen electrically connected with a button under the control of the observer.

Chronometer. A timepiece carefully constructed so as to keep accurate time.

Circinus (the Compass). One of the southern constellations.

Circle, Great. A circle on a sphere, the plane of which passes through the centre of the sphere.

Circle of Perpetual Apparition. A " small circle " of the celestial sphere, within which the stars do not pass below the horizon at any time. The radius of this circle is equal to the latitude of the place of observation. At the terrestrial poles, therefore, all the visible stars are within the circle of perpetual apparition, which is evidently bounded by the horizon. At the terrestrial equator there is no circle of perpetual apparition.

Circle of Position. A small circle on the earth's surface, the angular radius of which is equal to the sun's zenith distance at any given time. It is used in determining a ship's position at sea by Captain Sumner's method.

Circle, Small. Circles on a sphere, the plane of which does not pass through the centre of the sphere.

Circle, Transit. See TRANSIT INSTRUMENT.

Circles of the Celestial Sphere. Imaginary circles drawn on the celestial sphere, and used for purposes of astronomical measurement.

Circumpolar Stars. Stars which never set at the place of observation. The polar distance of such stars must, therefore, be less than the latitude of the place. It follows that at the terrestrial poles all the visible stars are circumpolar, and at the terrestrial equator there are no " circumpolar stars."

Civil Year. Usually 365 days, but once in every four years 366 days long (leap year). The average length of the civil year is nearly the same as that of the " tropical year," but 11¼ minutes longer. See LEAP YEAR and TROPICAL YEAR.

Clamp. A screw for temporarily tightening portions of astronomical instruments.

Clepsydra. An instrument for measuring time, used by ancient Greeks and Romans, and other nations. It consisted of a vessel filled with water, having a small hole in the bottom. The quantity of water discharged measured the lapse of time. Clepsydræ seem to have been invented by Ctesibius, of Alexandria, about 250 B.C.

Clock, Astronomical. See ASTRONOMICAL CLOCK.

Clock Stars. Stars used for finding the error of an astronomical clock.

Clusters, Star. Groups of small stars very close together. These are usually divided into (1) large and scattered clusters, (2) small compressed clusters, and (3) globular clusters.

Co-latitude. The trigonometrical complement of the latitude, or the difference between the latitude and 90°.

Collimating Eyepiece. An eyepiece used in the adjustments of a transit instrument.

Collimation, Error of. The line of collimation (which see) should be at right angles to the axis round which the telescope turns. If this be not so, the error is called the *error of collimation*.

Collimation, Line of. The line joining the optical centre of the object-glass of an astronomical telescope with the intersection of the middle wires in the eyepiece.

Collimators. Small telescopes placed due north and south of a transit instrument, and used for adjusting the line of collimation in the larger instrument.

Coloured Stars. Most of the stars are of different colours. Some are white or bluish-white, some yellow, others orange, and various shades of red (for list of remarkably red stars see Appendix). The components of many double stars show beautifully contrasted colours.

Columba (the Dove). One of the southern constellations. It lies south of Lepus, and south-west of Canis Major.

Colure, Equinoctial. The great circle, or circle of declination, which passes through the equinoctial points and the celestial poles.

Colure, Solstitial. The great circle which passes through the solstitial points and the celestial poles. This circle also passes through the pole of the ecliptic.

Comes. The fainter of the two components of a double star. Plural, *comites*.

Cometography. The department of astronomy which deals with comets.

Comets. " The word ' comet ' is derived through the Latin *cometa* and the French *comète* from the Greek κομήτης. In that language κόμη signifies the hair of the head ; and the first idea of comets was that they were bodies with hair-like appendages, appearing to stream from them like the hair from a person's head " (Lynn, *Remarkable Comets*). Some telescopic comets however, have no tails. Comets usually move round the sun in very elongated orbits. Some of them are periodical and therefore regular members of the solar system ; others are seen only once, and never return to the sun's vicinity.

Commensurability. A term applied to the equality between a certain number of periods of revolution of a planet or satellite with some other number of periods of another planet or satellite. Thus, two periods of revolution of Saturn round the sun are nearly equal to five of Jupiter. An example of commensurability is also found in the satellites of Saturn, the period of Tethys being double that of Mimas, and the period of Dione double that of Enceladus.

Commutation, Angle of. The angular distance between the sun's place, as seen from the earth, and that of a planet reduced to the ecliptic.

Compass, Points of. See Points of Compass.

Complement of an Angle. The difference between the angle and 90°.

Compression of a Planet. The amount by which a planet is flattened—like the earth—at the poles. This is usually expressed as follows : If e be the equatorial diameter, and p the polar, then the compression = $\frac{e-p}{e}$. For the earth the compression is about $\frac{1}{300}$; but for Jupiter, Saturn, and probably Uranus, the compression is considerably greater. The compression is also termed the *ellipticity* of the planet.

Cone. A solid which may be supposed to be generated by the revolution of a right-angled triangle round the perpendicular or vertical side of the triangle. The solid thus formed is termed a *right cone*, and the perpendicular of the generating triangle is called the axis. If the axis is not at right angles to the base, it is called an *oblique cone.*

Configuration. A term applied to the particular arrangement of the stars in a constellation or cluster, or the relative positions of the moon, planets, or other celestial bodies.

Conic Sections. These curves are known as the parabola, ellipse, and hyperbola. They are called conic sections because they may be supposed to be formed by the intersection of a plane and a cone. If the cutting plane intersects both sides of the cone, but is inclined to the axis of the cone, the section will be an ellipse; if it cuts obliquely parallel to the side of the cone, the boundary of the section will be a parabola; and if it be perpendicular to the base of the cone, the section will be an hyperbola. If the plane be perpendicular to the axis, the section will be a circle; so that the circle is merely a special form of the ellipse.

Conjunction. When two celestial bodies have the same longitude they are said to be in *conjunction.* The inferior planets Mercury and Venus are said to be in *inferior conjunction* with the sun when they pass between (or nearly between) the earth and sun. When in that part of their orbit which lies beyond the sun they are said to be in *superior conjunction.*

Co-ordinates. In analytical geometry the position of a point on a plane is determined by means of two co-ordinates. These may be either measured along two axes at right angles, when they are termed *rectangular co-ordinates,* or by an angle and distance, when they are called *polar co-ordinates.* The position of a point on the earth's surface is fixed by two co-ordinates, latitude and longitude. The position of objects on the celestial sphere are determined by the following systems: (1) Altitude and Azimuth; (2) North Polar distance and Hour Angle; (3) Right Ascension and Declination; (4) Latitude and Longitude (with reference to the ecliptic). (1) and (2) are affected by the earth's rotation; (3) and (4) are unaffected.

Constant. In astronomical and mathematical calculations a quantity which has always the same value.

Constellations. The groups or divisions into which the stars are divided for purposes of identification.

Copernican Theory. The theory that the sun forms the centre of the solar system. First advanced by Copernicus in the sixteenth century, and now universally accepted.

Cor Caroli. A name sometimes applied to the star α in Canes Venatici.

Cor Hydræ. A name sometimes applied to the star α Hydræ, otherwise known as Alphard.

Cor Leonis. A name sometimes applied to the star α Leonis or Regulus.

Corona Australis. One of the southern constellations.

Corona Borealis. One of the northern constellations.

Cor Serpentis. A name sometimes applied to the star α Serpentis.

Corvus (the Crow). One of the southern constellations.

Cosmical. A term applied to any fact or phenomenon connected with the heavenly bodies.

Co-tidal Lines. Imaginary lines on the earth's surface, passing through places where the tidal conditions are the same at the same time.

Crater (the Cup). One of the southern constellations.

Craters, Lunar. The ring-shaped formations on the moon's surface, visible with a telescope.

Crepuscular. A term applied to the twilight illumination of the sky.

Crux (the Cross). One of the southern constellations, containing the famous Southern Cross.

Culmination. The transit of a celestial body across the meridian of the place of observation. It then attains its highest altitude above the horizon ; hence the name.

Cursa. A name sometimes applied to the star β Eridani. Derived from the Arabic *kursi al-djauzâ al-mukaddam*, " the front throne of the giant," a term given by the Arabian astronomers to the stars λ, β, ψ Eridani and τ Orionis, which form a small quadrilateral figure close to Rigel.

Curtate Distance. The distance of a celestial body belonging to the solar system from the earth or sun when its place is projected on the plane of the ecliptic.

Cusps. The points of the " horns " of the crescent moon, or of the illuminated portion of the discs of Mercury and Venus when in the crescent phase.

Cycle. A period of time in which a series of celestial phenomena occur over again.

Cycle of Eclipses. A period during which eclipses of the sun and moon occur in nearly the same order. See SAROS and METONIC CYCLE.

Cygnus (the Swan). One of the northern constellations. Marked by the long cross formed by the stars a, δ, γ, ϵ, and β Cygni.

Cynosura. A name sometimes applied to the pole star (Polaris).

D.

Dark Glasses are used in telescopes when observing the sun. They are placed over the eyepiece to moderate the excessive glare and heat.

Day, Apparent Solar. The interval of time between one apparent noon and the next, or between two successive midnights.

Day, Lunar. The time which elapses between two successive passages of the moon across the meridian. The mean length of the lunar day is about 24 hrs. 50 mins. 32 secs.

Day Numbers, Bessel's. Small corrections to be added to the right ascensions and declinations of stars given for any certain epoch to reduce them to another epoch. These corrections are necessary to allow for the effects of precession, nutation, and aberration.

Day, Sidereal. The period of the apparent revolution of the stars round the celestial pole with reference to the meridian. The sidereal day is counted from *sidereal noon*, or the time of transit of the " First Point of Aries " across the meridian.

Declination denotes the angular distance of a celestial body north or south of the celestial equator. It is measured on a great circle passing through the body and the celestial pole. When the body is north of the equator, the declination is usually designated +, and when south of the equator, —.

Declination Circle. A great circle of the sphere passing through the celestial pole. On these circles the declinations of celestial bodies are measured. The term is also applied to the graduated circle of an equatorial telescope on which the declination of celestial objects is measured.

Declination Parallel. A small circle of the celestial sphere, every point on which has the same declination. The planes of these circles are consequently parallel to the plane of the equator.

Degree. In measuring angles the circle is divided into 360 equal parts. Each of these divisions is called a *degree.* The degree is subdivided into 60 minutes, and each minute into 60 seconds.

Deimos. The outer satellite of Mars, or that farthest from the planet. Its distance from the centre of Mars is about 14,500 miles, and it revolves round the planet in about 30 hrs. 18 mins. Its diameter is probably not more than 7 miles. Deimos was discovered by Professor Asaph Hall on Aug. 11th, 1877.

Delisle's Method of determining the Solar Parallax. In this method of observing transits of Venus, the sun's parallax is obtained by noting the difference between the times of beginning or ending of the transit from stations widely separated on the earth's surface. The places of observations must be near the earth's equator. Delisle's method succeeds best when the transit is nearly central—that is, when Venus passes nearly along a diameter of the sun's disc.

Delphinus (the Dolphin). One of the northern constellations, marked by a small rhomb of stars of fourth to fifth magnitude.

Deneb. A name sometimes applied to the star β Leonis.

Deneb Adige. A name applied to the star α Cygni. From the Arabic *dzanab al-dadjâdja*, "the tail of the hen" (or swan).

Deneb Algiedi. A name sometimes applied to the star δ Capricorni.

Denebola. A name sometimes applied to the star β Leonis. It is also called Deneb and Deneb Aleat.

Densities of Sun and Planets. From the principles of mathematical astronomy, the mass of the sun and planets can be found in terms of the mass of the earth. Then the density of the earth being known, and the relative volumes of the sun and earth, we can find the sun's density, or its specific gravity, with reference to an equal volume of water. In the same way the densities of the planets can be determined.

Density of the Earth. The relation between the weight of the earth as a whole and that of an equal volume of water. Taking water as 1, experiments made to determine the earth's density vary from 4·71 to 6·56. The result found by Francis Baily, by means of the " Cavendish Experiment "—namely, 5·66—is probably the best.

Descending Node of a planet's orbit (or comet's orbit) is the point where the planet's orbit cuts the ecliptic, when the planet is descending from the northern to the southern side of the ecliptic.

Diagonal Eyepiece. An eyepiece used in refracting telescopes for observing objects near the zenith. The rays from the object are reflected at right angles to the tube of the telescope by means of a prism or plane mirror.

Diameter, Apparent. The angle which the diameter of a celestial body subtends as viewed from the earth.

Dichotomy. A cutting in two. A term applied to the moon, Mercury, and Venus, when the illuminated portion is an exact semicircle.

Differentiation. The determination of the place of a celestial body by measurements from another the position of which is accurately known.

Digit (or Finger). A term used with reference to eclipses of the sun and moon. It denotes the $\frac{1}{12}$th of the diameter, and the number of digits indicates the magnitude of the eclipse.

Diminution of Gravity. The diminution in the weight of a body on the surface of the earth, or on the planets, due to the centrifugal force produced by the rotation of the earth or planet on its axis. This diminution attains its maximum at the equator.

Dione. One of the satellites of Saturn, the fourth in order counting from the planet, round which it revolves at a mean distance of about 239,000 miles in a period of 2 days 17 hrs. 41 mins. Its diameter is uncertain; but its stellar magnitude is, according to Professor Pickering, 11·5. It was discovered by J. D. Cassini in March 1684.

Dionysian Period. A period of 532 years, formed by multiplying together the lunar and solar cycles ($19 \times 28 = 532$). At the end of this period the moon's changes "take place on the same days of the week and month as before" (Chambers' *Descriptive Astronomy*).

Dioptrics. A division of the science of optics which deals with images formed by refraction through lenses.

Diphda. A name sometimes applied to the star β Ceti. From the Arabic *al-dhifda*, "the frog" (!).

Dip of Horizon. The angle between the horizontal line through the eye of an observer, and the line from his eye to the *offing*, or visible horizon, is called the *dip of the horizon*. This "dip" is due to the earth's rotundity, and increases with the height of the eye above the sea-level. The dip is diminished by refraction, which apparently increases the distance of the visible horizon.

Dip Sector. An instrument on the principle of double reflection, devised by Troughton for the determination of refraction, but subsequently used by Dr. Wollaston for measuring the dip of the horizon ; hence its name.

Direct Motion. The motion of a planet when it is moving from west to east among the stars. The term is also applied to the motions of comets when they move in the same direction as the planets, or contrary to the hands of a clock. It is also applied to the angular motion of the components of a binary star when the position angle is increasing.

Disc. The visible surface of the sun, moon, planets, and satellites.

Dispersion of Light. When a beam of white light is passed through a prism it is lengthened out into a rainbow-tinted band. This is due to the different rays being refracted in different degrees, and is called *dispersion*.

Disturbing Forces. Forces which tend to disturb the exact elliptical motion of a body round a centre of force. Thus, the moon's motion round the earth is disturbed by the attraction of the sun and planets, and the motion of the earth and planets is disturbed by the attraction of each other.

Diurnal Aberration. See ABERRATION.

Diurnal Libration. A small libration of the moon, due to the earth's rotation on its axis. When the moon is rising we see a little more of its western side than when it is near the zenith, and when it is setting a little more of the eastern side. The effect is really due to parallax, caused by the earth's rotation on its axis. Its maximum amount is equal to the moon's horizontal parallax, or about 57'.

Diurnal Motion. The apparent motion of the celestial bodies from east to west, due to the rotation of the earth on its axis from west to east.

Dorado (the Sword-Fish). One of the southern constellations.

Double Stars. Stars which appear as single stars to the naked eye, but are seen to consist of two stars when viewed with a telescope. Some of these objects are visible with small telescopes, while others have their components so close that they require the largest instruments to divide them. If the components revolve round each other the double star is called a *binary star*. If there is no relative orbital motion the object is called an *optical double*, as one component may possibly be far out in space beyond the other, and only accidentally placed nearly in the same direction.

Draco (the Dragon). One of the northern constellations.

Draconids. A meteor shower visible about August 21st to 23rd in each year. The meteors seem to radiate from a point in the constellation Draco (291° + 60°). They are slow, with trains.

Dubhe. A name sometimes applied to the star α Ursæ Majoris, the northern of the two " pointers."

Dynamical Mean Sun. An imaginary sun, or rather point, which is supposed to coincide with the true sun at perigee and to move along the ecliptic at a mean rate in a period of one year.

Dynamometer. An instrument for measuring the magnifying power of the eyepieces of telescopes.

E.

Earth. The planet which we inhabit. The earth is an oblate spheroid—that is, it is slightly flattened at the north and south poles. For data respecting the earth, see Appendix.

Earth Shine. The dark part of the moon visible a little before and a little after "new moon." It is due to reflected light from the earth; hence the term. It is called by the French *Lumière cendrée*.

Earth's Way. The angle between the direction in which a star is seen and the direction of the earth's orbital motion at the time. It is used in calculating the *coefficient of aberration*. See ABERRATION.

Eccentricity of an Orbit. In an elliptic orbit the distance of each of the foci from the centre of the ellipse. The eccentricity is usually expressed as a decimal fraction of the semi-axis major of the ellipse. Thus, if the eccentricity of an orbit be 0·20, it means that each of the foci lie at a distance from the centre equal to $\frac{1}{5}$th of the semi-axis major.

Eccentricity of the Earth's Orbit. The eccentricity is at present about 0·01677, or nearly $\frac{1}{60}$. According to Le Verrier, the eccentricity varies between the limits 0·0747 and 0·0047. At present the eccentricity is diminishing, but will not reach its minimum value for many thousand years. Harkness gives the formula

$$c = 0{\cdot}016771049 - 0{\cdot}000,0004245\,(t - 1850) - 0{\cdot}000,000,001367 \left(\frac{t - 1850}{100}\right)^{2}$$

for the eccentricity at any future epoch, t.

Eclipse. The passing of one celestial body through the shadow of another, as the passage of the moon through the earth's shadow, the disappearance of the satellites of Jupiter in the shadow of the planet, etc.

The term is also usually applied to eclipses of the sun; but these are more correctly *occultations* by the moon (see OCCULTATION). A true eclipse is one in which the surface of the body is actually darkened; but in the case of solar eclipses the sun's surface is evidently not darkened.

Ecliptic. The great circle of the celestial sphere along which the sun apparently travels during the year.

Ecliptic, Obliquity of. See OBLIQUITY.

Egress. The end of a transit of Mercury or Venus, when the planet passes off the sun's disc, or of a satellite off the disc of its primary.

Electra. One of the stars in the Pleiades.

Elements of a Binary Star Orbit. Quantities which define the position of the stellar orbit in space with reference to a tangent plane to the celestial sphere (or the background of the sky) at the place of the primary star. Also the time of revolution of one component round the other, or rather of both round the common centre of gravity, the epoch of the periastron passage, and the eccentricity of the real orbit. These elements are: P, the period in years; T, the time of periastron passage; e, the eccentricity of the real orbit; Ω, the position angle of the line of nodes; i or γ, the inclination of the orbit to the plane of projection; λ, the position of the periastron measured from the node on the real orbit; and a, the semi-axis major of the real orbit in seconds of arc.

Elements of a Comet's Orbit. Quantities which define the position of a comet's orbit in space with reference to the plane of the ecliptic. These are: π, the longitude of the perihelion, or the comet's longitude

when it passes through that point; Ω, the longitude of
the ascending node as seen from the sun; q, the perihelion
distance from the sun, expressed in terms of the earth's
mean distance from the sun; i, the inclination of the
orbit to the plane of the ecliptic.   Other elements
are: the time of perihelion passage, and—if the orbit is
elliptic—the period of revolution round the sun in years,
and the eccentricity of the orbit.

Elements of a Planet's Orbit.   Quantities which
determine the position of a planet's orbit in space with
reference to the plane of the ecliptic.   These are: a, the
mean distance from the sun (that of the earth $= 1$);
P, the mean sidereal period in mean solar days; e, the
eccentricity; i, the inclination of the orbit plane to the
plane of the ecliptic; Ω, the longitude of the ascending
node; π, the longitude of the perihelion measured from
the node on the orbit; L, the mean longitude of the
planet at a certain time; and E, the epoch for which
L is given.

Elements of a Variable Star.   These are: (1)
epoch of maximum or minimum light; (2) the mean
length of the period from maximum to maximum, or
from minimum to minimum; (3) the variation of light,
or the stellar magnitude at maximum and the magnitude
at minimum.   In some cases more elaborate formulæ
are given.

Elevation.   A term sometimes applied to the altitude
of a celestial body above the horizon.   See ALTITUDE.

Ellipse.   One of the conic sections.   An ellipse may be
supposed to be formed by the intersection of a plane with
a cone, the cutting plane being inclined to the axis of
the cone, and cutting both sides of the cone.   In the
ellipse, the distance of every point on the curve from

a fixed point within it is proportional to its perpendicular distance from a fixed line outside the curve. The fixed point is called the *focus*, and the fixed line the *directrix*. There are two *foci*, both situated on the longer axis of the ellipse. Another property of the ellipse is that the sum of the distances of any point on the curve from the foci is constant and equal to the major axis of the ellipse.

Elliptic Motion. When one body revolves in an elliptic orbit round another situated in one of the foci of the ellipse the motion is called *elliptic motion*.

Ellipticity of the Earth and Planets. Same as COMPRESSION, which see.

Elongation. The difference between the celestial longitude of a planet and that of the sun. The elongation of a satellite is the angular distance of the satellite from its primary.

Emersion. The reappearance of a star or planet after occultation by the moon, or the reappearance of a satellite after being eclipsed in the shadow of its primary.

Enceladus. One of the satellites of Saturn, the second in order counting from the planet, round which it revolves in a period of 1 day 8 hrs. 53 mins. at a mean distance of about 151,000 miles. Its diameter is uncertain. It was discovered by Sir William Herschel on August 28th, 1789. According to Professor Pickering, the stellar magnitude of Enceladus at mean opposition is 12·3.

Enib. A name sometimes applied to the star ε Pegasi.

Epact. A number employed in the construction of the calendar.

Ephemeris. A table showing the predicted positions of a moving celestial body.

Epicycle. A small circle the centre of which lies on the circumference of a larger circle. It was a device used by the ancient astronomers to explain the apparent motions of the planets, when the earth was supposed to be the centre of the planetary system.

Epoch. A date of reference used in astronomical calculations.

Equation, Annual. An inequality in the moon's motion due to the varying distance of the sun from the earth.

Equation of Equinoxes. Owing to nutation there is a periodical oscillation of the " First Point of Aries," the period being about $18\frac{2}{3}$ years. The angular distance between the mean and true position is called the *Equation of the Equinoxes*, and is about 15′ 37″.

Equation of Light. The time taken by light to pass from the sun to the earth. This is about 8 mins. 18 secs.

Equation of the Centre. The angle by which the true longitude of the earth differs from its mean longitude. Its maximum value is 1° 55′ 33·3″. The term is also applied in the same sense to the orbits of the planets.

Equation of Time. The amount which it is necessary to add to or subtract from the apparent time in order to obtain the mean time.

Equation, Personal. The error in the time of transit of a celestial body by a particular observer is called his " personal equation." The term might also be applied to other observations, such as the relative brightness of white and coloured stars, etc.

Equations of Condition. Equations which express the relations existing between the coefficients of another equation. These equations are employed to determine from observations the values of the coefficients in a general equation. They are usually solved by a method known as the Method of Least Squares, which see.

Equator, Celestial. The great circle in which the plane of the terrestrial equator (produced) intersects the star sphere. Every point on the celestial equator is 90° distant from either celestial pole.

Equator of a Planet. The great circle on the surface of the planet the plane of which is at right angles to the planet's axis of rotation.

Equator, Terrestrial. The great circle on the earth's surface every point on which is equidistant from either pole. The plane of the equator is perpendicular to the earth's axis.

Equatorial Horizontal Parallax. The geocentric parallax of a celestial body, as viewed from a place on the earth's equator. It is therefore the angle whose sine is the equatorial radius of the earth divided by the distance of the body from the earth's centre.

Equatorial Telescope. A telescope mounted with its principal axis pointing to the celestial pole. This axis is therefore parallel to the earth's axis of rotation. Perpendicular to the polar axis is a secondary one, which carries the telescope at right angles to it. The secondary axis is movable on the primary so that the telescope may be pointed to any star. A star may thus be kept in view by one motion. An equatorial telescope is usually fitted with graduated circles, and the motion necessary to follow a star is communicated by clockwork.

Equinoctial Points. The points at which the equator and the ecliptic intersect each other. One of these is called the "First Point of Aries," and is denoted by the symbol ♈. The opposite point is called the "First Point of Libra," and is denoted by ♎. Owing to the "precession of the equinoxes" the former has now retrograded into Pisces, and the latter into Virgo. See PRECESSION OF EQUINOXES.

Equinoxes. The points at which the ecliptic intersects the plane of the celestial equator. The point at which the sun, moving in the ecliptic, passes from the south to the north of the equator is called the *Vernal Equinox*, and the point at which the sun passes from the north to the south of the equator the *Autumnal Equinox*. This of course applies only to the earth's northern hemisphere; for in the southern hemisphere the above terms would be reversed.

Equinoxes, Precession of. See PRECESSION OF EQUINOXES.

Equuleus (the Little Horse). One of the northern constellations.

Eridanus (the River). One of the southern constellations.

Errai. A name sometimes applied to the star γ Cephii.

Error, Probable. A term used with reference to a series of observations, each of which is subject to a small error. "In any series of errors the probable error has such a value that the number of errors greater than it is the same as the number less than it; or it is an even wager that an error taken at random will be greater or less than the probable error" (Professor Merriman, *Method of Least Squares*, p. 66).

Establishment of the Port. The time which elapses after the moon's transit across the meridian before high water occurs at the given port. This is for the "lunar tide." A similar correction must be made for the "solar tide."

Etanin. A name sometimes applied to the star γ Draconis. From the Arabic *ras al-tannin*, "the dragon's head."

Ether. The supposed medium which pervades all space, and through which, by means of wave motion, light and heat, and perhaps electricity, are transmitted from the sun and stars to the earth.

Evection of Moon. An inequality in the moon's motion, due to the elliptical shape of its orbit. The sun's disturbing force produces periodical changes in the eccentricity depending on the position of the line of apsides, and these changes are termed the *evection*. The inequality was discovered by Ptolemy, but was previously suspected by Hipparchus.

Exterior Planets. The planets which revolve round the sun at a greater distance than the earth. These are Mars, the Minor Planets or Asteroids, Jupiter, Saturn, Uranus, and Neptune.

Extinction of Light. A supposed diminution in the brightness of the stars as seen from the earth, due to an absorption of light in the luminiferous ether of space. That such an absorption of light does take place in the ether has not, however, been well established.

Eyepiece. The lens or combination of lenses placed at the eye end of a telescope. Its object is to magnify he image formed by the object-glass.

F.

Faculæ of the Sun. Brighter portions of the sun's surface, usually seen near sun-spots, or in places where spots have disappeared or are about to appear.

Falcated. A term applied to the Moon, Mercury, and Venus when they are in the crescent phase.

Field of View. The portion of the sky visible in a telescope. With high magnifying powers the "field of view" is much smaller than with low powers.

Filar Micrometer. A form of micrometer in which fine wires are used.

First Quarter. See QUARTER.

Flat. The small plane mirror used in the Newtonian form of reflecting telescope to reflect the rays from the large mirror into the eyepiece.

Flora. One of the minor planets revolving round the sun in orbits lying between those of Mars and Jupiter. It was discovered by Hind on Oct. 18th, 1847. It revolves round the sun in a period of 3·266 years at a mean distance of 2·20 times the sun's mean distance from the earth. Its opposition magnitude is about 9, and its real diameter perhaps about 60 miles.

Foca. A name sometimes applied to the star α Coronæ Borealis.

Foci of an Ellipse. Two points on the longer axis of an ellipse equidistant from the centre. The distance of each focus from the centre depends upon the "eccentricity" of the ellipse. The distance of each focus from either extremity of the minor axis is equal to the semi-axis major; and the sum of the distances of any point on the curve from the foci is constant, and equal to the major axis.

Focus. The point in which rays of light unite after undergoing refraction through lenses or reflection from a mirror.

Fomalhaut. A name applied to the bright southern

star *a* Piscis Australis. From the Arabic *fum al-hût al-djanûbi*, " the mouth of the southern fish."

Forces, Disturbing. See DISTURBING FORCES.

Fornax (the Furnace). One of the southern constellations.

Foucault's Experiment. An experiment devised by Foucault to render the earth's rotation on its axis visible to the eye. A heavy metal ball suspended by a long and fine wire is set vibrating like a pendulum. It will be found that the plane of vibration apparently rotates from east to west, or contrary to that of the earth's rotation. At the poles the plane of vibration would make one rotation in a sidereal day. At places between the poles and the equator the time of rotation depends on the latitude of the place of observation. At the equator there would be no rotation.

Fraunhofer's Lines. The dark lines seen in the spectra of the sun, moon, planets, and fixed stars. They were discovered by the famous German optician Fraunhofer ; hence their name.

Front View. A form of reflecting telescope devised by Sir William Herschel. There was no small mirror, but the image formed by the large mirror was (by slightly shifting the position of the large mirror) thrown to the side of the tube, where it was examined directly by the eyepiece. See HERSCHELIAN TELESCOPE.

Full Moon. When the moon is in opposition to the sun, or distant from it by 180° of celestial longitude, it is said to be " full." Accurately speaking, however, the moon is not truly " full " except during the totality of a lunar eclipse.

G.

Galactic Circle. A term applied to the mean or centre line of the Galaxy, or Milky Way zone.

Galaxy. Another name for the Milky Way, which see.

Gauges, Star. A term applied by Sir William Herschel to his counts of stars visible in the field of his telescope in various parts of the sky.

Gegenschein. A phenomenon connected with the zodiacal light. It is a small spot of faint light seen in the sky opposite to the sun's place—that is, 180° from the sun. Keen eyesight is necessary for its detection, as it is always very faint.

Gemini (the Twins). One of the zodiacal constellations. Its brightest stars are Castor and Pollux.

Geminids. A meteor shower visible about December 9th to 12th in each year. The meteors seem to radiate from a point near Castor (107° + 33°). They are swift, with short paths.

Gemma. A name sometimes applied to the star α Coronæ Borealis.

Geocentric Latitude. The latitude of a celestial body as supposed to be seen from the centre of the earth.

Geocentric Longitude. The longitude of a celestial body as supposed to be seen from the earth's centre.

Geocentric Lunar Distances. The angle subtended between the centre of the moon's disc and a given star as seen from the earth's centre. It was formerly used in the determination of terrestrial longitude at sea; but chronometers are now more generally relied on for this purpose.

Geocentric Parallax. The angle subtended at a celestial body by the earth's radius at the point of observation. Hence, to find the geocentric place of the body, or its position as seen from the earth's centre, the amount of the geocentric parallax must be deducted from the apparent zenith distance. The "fixed stars" have no geocentric parallax.

Geocentric Place. The position of a celestial body as supposed to be seen from the earth's centre.

Geodesy. The science which treats of the figure and dimensions of the earth.

Giauzar. A name sometimes applied to the star λ Draconis.

Gibbosity of Mars. When Mars is between " opposition " and " quadrature " (or 90° from the sun) it is slightly gibbous, like the moon a little before and a little after " full moon." At quadrature about one-eighth of the planet's disc is in shadow. For Jupiter and the other planets exterior to Mars the gibbosity is not perceptible.

Gibbous Moon. The moon's phase when more than half the disc is illuminated. This occurs between " first quarter " and "full moon" and again between "full moon" and " last quarter."

Giedi, Prima and Secunda. Names sometimes applied to the stars a^2 and a^1 Capricorni, which form a double star to the naked eye.

Gjenula. A name sometimes applied to the star γ Aquarii.

Globe, Celestial. A globe showing the positions of the stars on the celestial sphere. The observer's eye is supposed to be placed at the *centre of the globe.* It follows therefore that the constellations as marked on a celestial globe are inverted, and are not in their true configurations as seen in the sky.

Globular Clusters. A name applied to close star clusters of a spherical or nearly spherical form.

Gnomon. Another name for a sun-dial. Derived from the Greek γνώμων, an index.

Golden Number. The remainder when the number of the year increased by one is divided by 19. Thus the golden number for 1893 is the remainder when 1894 is divided by 19—that is, 13. If exactly divisible by 19, then 19 is the golden number. Thus the golden number for 1899 will be 19.

Gomeisa. A name sometimes applied to the star β Canis Minoris.

Granulation. A term applied to the mottled appearance visible through a telescope on the sun's surface. The granulations have also been called "willow leaves" and "rice grains."

Gravitation. The tendency of all bodies in the universe to attract each other. The phenomenon of terrestrial gravity has, of course, been known for ages, but the laws of universal gravitation were discovered by Sir Isaac Newton.

Great Circle. The circle on a sphere, the plane of which passes through the centre of the sphere.

Gregorian Reform of Calendar. See CALENDAR, GREGORIAN.

Gregorian Telescope. A form of reflecting telescope in which the smaller mirror is concave, and the reflected rays from the object pass through a circular opening in the large mirror.

Grummium. A name sometimes applied to the star ξ Draconis.

Grus (the Crane). One of the southern constellations.

Gyroscope. A spinning top, or heavy rotating wheel, with its axis of rotation supported in a ring, which is again supported in another ring. This second ring rotates in a fixed frame. By this arrangement the axis of the wheel can be made to point in any direction. The instrument is used to illustrate the earth's rotation on its axis, and the precession of the equinoxes.

H.

Hadley's Sextant. A form of the sextant invented by John Hadley in 1730. An instrument of almost the same form was invented by Thomas Godfrey in the same year. An improved form of the instrument, very similar to that now in use, was devised by Hadley shortly after his first invention. Hadley and Godfrey each received £200 from the Royal Society for their invention. A similar instrument is said to have been invented by Sir Isaac Newton. See SEXTANT.

Halley's Comet. One of the periodical comets, or comets which revolve round the sun and regularly return. The period of Halley's comet is about seventy-six years. Various apparitions of this comet have been traced back, the earliest recorded being in the year B.C. 44. It appeared last in 1835, and its next return will be due about 1910.

Halley's Method of Determining the Solar Parallax. This method of finding the sun's parallax from observations of transits of Venus was devised by Halley in 1716. Two stations are selected—one in high northern latitudes, and the other in high southern latitudes, and both lying as nearly as practicable in a plane at right angles to the orbit plane of Venus. The times of duration of the transit are observed from the two stations, and from these durations the solar parallax is computed.

Hamal. A name sometimes applied to the star a Arietis.

Harmonic Circle. If chords be drawn through the focus of an ellipse, and harmonic means be taken between the intercepts from the focus to the curve, these harmonic means, when laid off from the focus on the chord, will give a number of points which all lie on a circle of which the focus is the centre and the diameter the *latus rectum* of the ellipse (see LATUS RECTUM). This circle is called the " harmonic circle," and is used in the calculation of the orbit of a binary star by the graphical method.

Harmonic Ellipse. The ellipse into which the harmonic circle is projected in the apparent orbit of a binary star.

Harvest Moon. The full moon which falls nearest to the autumnal equinox, or Sept. 23rd, in each year. At this time the interval of time between moonrise on successive nights is much smaller than usual, owing to the fact that the ecliptic then makes the smallest angle with the horizon.

Hebe. One of the group of minor planets which revolve round the sun in orbits lying between those of Mars and Jupiter. It was discovered by M. Heneke at Driesen on July 1st, 1847. When in opposition its stellar magnitude is about $8\frac{1}{2}$. Its period of revolution round the sun is about 3·776 years, and the eccentricity of its orbit about the same as that of Mercury (0·20).

Heliacal. A term applied to the rising or setting of a celestial body at the same time as the sun.

Heliocentric Place. The position of a celestial body as supposed to be seen from the centre of the sun.

Heliometer. An astronomical telescope in which the object-glass is cut in two along a diameter. One half is made to move along the other by a graduated screw. Each half, when separated, forms a distinct image of the object viewed. The instrument may be used for finding the diameters of the sun, moon, and planets, or the angular distance between the components of a double star.

Helioscope. A little instrument devised by Dawes for facilitating the telescopic observation of the sun's surface. It consists of a metallic plate, pierced with a minute hole, and placed in the focus of a telescope. By this means the eye is protected from the glare.

Heliostat. An instrument devised to reflect a ray of sunlight in a fixed direction. When used for astronomical purposes it is sometimes called a *siderostat*.

Hemisphere. The half of a sphere. A plane passing through the centre of a sphere divides it into two hemispheres.

Hercules. One of the northern constellations.

Herschelian Telescope, or **"Front View."** A form of reflecting telescope, devised by Sir William Herschel, in which the second mirror is dispensed with. The large mirror is inclined slightly to the axis of the tube, and the image formed is viewed directly by the eyepiece placed at the edge of the tube.

Homam. A name sometimes applied to the star ζ Pegasi.

Horary. A term applied to phenomena connected with an hour.

Horizon, Celestial. The tangent plane to the surface of the earth at the place of observation, and produced to meet the star sphere. In other words, it is a plane perpendicular to the diameter of the earth at the observer's standpoint. The poles of this plane are the zenith and nadir.

Horizon, Rational. A plane through the earth's centre parallel to the celestial or sensible horizon. It is sometimes termed "the true horizon."

Horizon, Sensible. Same as Celestial Horizon, which see.

Horizontal Parallax. The geocentric parallax of a celestial body when the body is on the horizon of the plane from which it is observed. It varies inversely as the distance of the body from the earth. The stars are at such a vast distance that their geocentric parallax is inappreciable. See GEOCENTRIC PARALLAX.

Horologium (the Clock). One of the southern constellations.

Hour Angle. The angle between a star's declination circle and the meridian of the place of observation.

Hour Circle. The graduated circle of an equatorial telescope on which the right ascensions of celestial bodies are measured.

Hunter's Moon. The full moon which falls nearest to Oct. 21st in each year. The phenomenon somewhat resembles the harvest moon, but is less marked.

Hyades. A remarkable group of stars shaped like a V in the constellation Taurus.

Hydra (the Sea Serpent). One of the constellations. It is of great length, extending over seven hours of right ascension.

Hydrus (the Water Snake). One of the southern constellations.

Hyperbola. One of the conic sections, which may be supposed to be formed by the intersection of a plane with a cone, the plane being perpendicular to the base of the cone.

Hyperion. One of the satellites of Saturn—the seventh in order counting from the planet. It was discovered by Messrs. Bond and Lassell, on Sept. 19th, 1848. Its mean distance from Saturn is about 951,000 miles, and its period of revolution round the planet 21 days 6 hrs. 39 min. According to Professor Pickering, the stellar magnitude of Hyperion at mean opposition is 13·7. Its real diameter is doubtful.

I.

Iklil, or **Iklil-al-Jebhah.** An Arabic name for the star β Scorpii.

Illumination of the Field of View. For the purpose of measuring celestial bodies, very fine wires are fitted in the eyepiece of a telescope. To render these visible on a dark sky it is necessary to use a small lamp, the light of which is thrown in through a hole in the tube. There are two methods of illumination used: (1) dark wires in a bright field, and (2) bright wires in a dark field.

Immersion. The disappearance of a star or planet when occulted by the moon, or the disappearance of a satellite in the shadow of its primary.

Inclination of Orbit. The angle between the plane of an orbit and a given plane of reference. The orbits of the planets and comets are referred to the plane of the ecliptic; the orbits of the binary stars to a tangent plane to the star sphere at the plane of the star, or in other words, to the background of the sky.

Indiction. A period of fifteen years, fixed by the Roman emperor Constantine as a conventional division of time.

Indus (the Indian). One of the southern constellations.

Inequality, Moon's Parallactic. An inequality in the moon's motion due to the varying amount of the sun's disturbing force at "new moon" and "full moon." It tends to accelerate the time of "first quarter" and to retard that of "last quarter."

Inequality of Jupiter and Saturn. An inequality in the orbital motion of these large planets due to their attractions on each other. It depends upon the near commensurability of the periods of revolution of Jupiter and Saturn round the sun, two periods of Saturn being nearly equal to five of Jupiter.

Inferior Conjunction. See CONJUNCTION.

Inferior Planet. A planet which revolves round the sun at a distance less than that of the earth. There are only two inferior planets—Mercury and Venus.

Ingress. The beginning of a transit of Mercury or Venus, when the planet passes on to the sun's disc. The term is also used with reference to the transits of the satellites of Jupiter and Saturn across the disc of their primary.

Instruments, Meridian. Instruments used in observations for observing the stars when crossing the meridian.

Intercalation. A term applied to the addition of one day to each leap year, or every fourth year. These added days are termed *intercalary* or *leap days*.

Interior Planets. Same as Inferior Planets, which see.

Interpolating Curve. A term applied to a curve drawn through a number of observations (such as the measures of a binary star) plotted on square ruled paper.

This curve should be a "smooth" one—that is, without sudden changes of curvature—and should leave as many observations on one side of the curve as on the other, or as nearly so as possible.

Interstellar. The portion of space which contains the fixed stars, or those parts which lie outside the solar system.

Izar. A name sometimes applied to the star ε Boötis.

J.

Japetus. The outer satellite of Saturn, or that farthest from the planet. It was discovered by J. D. Cassini, on Oct. 25th, 1671. Its mean distance from Saturn is about 2,261,000 miles, and its period of revolution 79 days 7 hrs. 54 mins. According to Professor Pickering, its stellar magnitude at mean opposition is 11·8.

Jovicentric. The place of a celestial body with reference to the centre of the planet Jupiter.

Julian Calendar. The introduction of a leap year every four years, due to Julius Cæsar, B.C. 44.

Julian Period. "A period useful in chronology is obtained by multiplying together the lunar cycle, the solar cycle, and the indiction, forming a period of 7980 years ($19 \times 28 \times 15 = 7980$)" (Chambers' *Handbook of Descriptive Astronomy*).

Juno. One of the minor planets revolving round the sun in orbits lying between those of Mars and Jupiter. It was discovered by Harding Sept. 1st, 1804. It revolves round the sun in a period of 4·358 years, at a mean distance of 2·668 times the earth's mean distance from the sun. When in opposition its magnitude is about 8·5, and its real diameter is perhaps about 120 miles.

Jupiter. The largest of all the planets of the solar system, its mean diameter being about 87,000 miles or about eleven times that of the earth. Its volume, therefore, exceeds that of the earth over 1300 times; but in density it is light, its mass being only 312 times the mass of the earth. Its mean distance from the sun is about 483,000,000 miles; and it revolves round the sun in a period of 11 years, 314·8 days. Jupiter has five satellites, the nearest and smallest having been discovered by Barnard in September 1892. For further details, see Appendix.

K.

Kaffaljidhma. A name applied by the Arabian astronomers to the star γ Ceti.

Kaitain. A name sometimes applied to the star a Piscium.

Kaus Australis. A name sometimes applied to the star ϵ Sagittarii.

Keid, or **Al-kaid.** A name applied by the Arabian astronomers to the star 40 (o^2) Eridani.

Kepler's Laws. Laws of planetary motion discovered by the famous Danish astronomer Kepler. These laws are as follows :—

I. The planets revolve round the sun in elliptic orbits, with the sun in one of the foci of the ellipse.

II. The radius vector, or straight line joining the centres of the sun and planet, sweeps over equal areas in equal times.

III. The squares of the periodical times of the different planets are proportional to the cubes of their mean distances from the sun.

Kepler's Laws also apply to the motions of the satellites round the planets, and, with suitable modifications, to the revolution of the components of binary stars round their common centre of gravity.

Kiffa Australis. A name sometimes applied to the stars a^1 and a^2 Libræ.

Kiffa Borealis. A name sometimes applied to the star β Libræ.

Known Stars. Stars whose position on the celestial sphere have been accurately determined by meridianal observations.

Kocab. A name sometimes applied to the star β Ursæ Minoris.

Korneforos. A name sometimes applied to the star β Herculis.

L.

Lacerta (the Lizard). One of the northern constellations. It lies between Cepheus and Cygnus.

Lady's Way. A name formerly applied to the zodiac.

Lagging of Tides. A delay in the time of high water which occurs between the " first quarter " and " full moon," and between the " last quarter " and " new moon," due to the combined action of the sun and moon.

Last Quarter of Moon. See QUARTERS.

Latitude, Celestial. The angular distance of a celestial body from the ecliptic, measured on a great circle at right angles to the ecliptic.

Latitude, Geocentric. The angular distance of a celestial body, north or south of the ecliptic, as supposed to be seen from the centre of the earth.

Latitude, Heliocentric. The angular distance of a celestial body, north or south of the ecliptic, as supposed to be seen from the centre of the sun.

Latitude, Parallel of. A "small circle" on the earth's surface parallel to the equator.

Latitude, Terrestrial. The angular distance of a place on the earth's surface, north or south of the terrestrial equator.

Latus Rectum, or Parameter. The chord drawn through the focus of a conic section at right angles to the major axis. If semi-axis major of an ellipse $= a$, eccentricity $= e$; then length of latus rectum $= 2a\,(1 - e^2)$.

Leap Year. The ordinary civil year consists of 365 days; but as the real period of the earth's revolution round the sun is about $365\frac{1}{4}$ days, a day is added to every fourth year, which has therefore 366 days, and is called Leap Year. As this correction is not exactly accurate, the leap year is omitted every hundred years, when the last year of the century is not divisible by 400. Thus, 1700 and 1800 were not leap years, and 1900 will not be a leap year; but the year 2000 will be a leap year.

Least Squares, Method of. A method of solving a number of equations of condition invented by Gauss. The method is as follows: Multiply each equation by the coefficient of the first term, and add. Multiply each equation by the coefficient of the second term, and add; and so on. We thus obtain as many equations as there are unknown quantities, and these can be solved by one of the usual methods of solving simultaneous equations.

Lemniscate. A term applied to the dark opening in the great nebula in Argo, which is sometimes spoken of as the "key-hole nebula."

Lens. "A portion of a refracting medium bounded by two spherical surfaces; the straight line joining their centres being called the *axis* of the lens" (Osmund Airy, *Geometrical Optics*). Lenses are usually formed of glass.

Leo (the Lion). One of the constellations of the zodiac. It contains the well-known "Sickle." Its brightest star is Regulus (a Leonis).

Leo Minor (the Lesser Lion). One of the northern constellations. It lies between Ursa Major and Leo.

Leonids. A meteor shower visible about Nov. 13th to 14th in each year. The meteors seem to radiate from a point near ζ Leonis ($149° + 33°$). They are very swift, with streaks. They are especially numerous once every thirty-three years, when a magnificent shower is usually visible. The last great shower of Leonids occurred in November 1866, and the next will be due in November 1899.

Lepus (the Hare). One of the southern constellations. It lies between Orion and Columba.

Libra (the Balance). One of the constellations of the zodiac.

Libration. The rotation of the moon on its axis is uniform, but its orbital motion round the earth is not so, owing to the elliptical shape of the orbit. This inequality between the velocities of rotation and revolution gives rise to an apparent oscillation of the moon's disc, which brings alternately into view small portions of the opposite hemisphere near the east and west limbs. This is called the *libration in longitude*. Another libration, called the *libration in latitude*, is due to the fact that the moon's axis of rotation is not exactly perpendicular to the plane of her orbit.

Libration, Diurnal. See DIURNAL LIBRATION.

Light Year. The distance which light travels in one year. The distance of stars from the earth is sometimes expressed by stating the number of years which light would take in passing from the star to the earth. When the "parallax" is known, the number of years' travel for light may be found by dividing the number 3·258 by the parallax expressed as a fraction of a second of arc.

Limb. The edge of the disc of the sun, moon, or planets.

Limits, Ecliptic. The angular distance from the node of the moon's orbit on the ecliptic within which an eclipse is possible. For solar eclipses the moon must be within 16° 58' of the node. For lunar eclipses the sun must be within 11° 21' of the moon's node in order that there may be any contact of the moon with the *umbra* of the earth's shadow.

Local Time. The mean time at any given place on the earth's surface.

Longitude, Celestial. The angular distance of a celestial body from the "First Point of Aries" *measured on the ecliptic*. This, combined with the latitude, fixes the position of a body on the celestial sphere. See LATITUDE.

Longitude, Geocentric. The longitude of a celestial body as supposed to be seen from the centre of the earth.

Longitude, Heliocentric. The longitude of a celestial body as supposed to be seen from the centre of the sun.

Longitude, Terrestrial. The angular distance of a place on the earth's surface east or west of a fixed meridian such as the meridian of Greenwich.

Longitude of Perihelion. The longitude of the perihelion of the orbit of a planet or comet as supposed to be seen from the centre of the sun. It is usually measured on the ecliptic to the node of the orbit, and from the node along the orbit to the perihelion point. A more satisfactory method, however, would be to state the heliocentric longitude of the perihelion point.

Loop of Retrogression. The loop in a planet's apparent path in the sky described when the planet's motion is changing from direct to retrograde and *vice versâ*.

Lucida. A term sometimes applied to the brightest object in a group of stars. Thus Alcyone may be called the *lucida* of the Pleiades.

Luculi. A term sometimes applied to the small bright spots visible on the sun's surface.

Lumière Cendrée. A term applied to the "earthshine" visible on the moon when in the crescent phase.

Lunar Cycle. Same as Meteoric Cycle, which see.

Lunar Distances. The angular distance of the moon's centre from the sun or from bright stars and planets which lie near its path in the sky. Tables of these computed distances are given in the Nautical Almanack for every third hour of Greenwich mean time.

Lunar Inequalities. Inequalities or deviations from a regular elliptic orbit produced in the moon's motion round the earth by the attraction of the sun and planets.

Lunation. Same as a Synodic Month, which see.

Lune. The crescent-shaped space contained between two intersecting circles.

Lupus (the Wolf). One of the southern constellations.

Lynx (the Lynx). One of the northern constellations.

Lyra (the Lyre). One of the northern constellations.
Its brightest star is the brilliant Vega (a Lyræ).

Lyrids. A shower of meteors visible about April
19th to 30th in each year. They seem to radiate from
a point near the constellation Lyra (271° + 33°).

M.

Maculæ. A term sometimes applied to the darker
portions of sun spots.

Magellanic Clouds. Two spots of nebulous light
visible to the naked eye in the southern hemisphere, and
distinct from the Milky Way. The larger is known to
astronomers as the *Nubecula Major*, and the smaller as the
Nubecula Minor. Both consist of a collection of small
stars, star clusters, and nebulæ.

Magnetic Storm. A disturbance in the magnetic
conditions of the earth, probably due to a disturbance in
the sun. A magnetic storm is indicated by large and
sudden variations in the magnetic needle, by auroras,
etc.

Maia. One of the stars in the Pleiades.

Major Axis of Orbit. In an elliptical orbit, the
longer axis, or that passing through the two foci, is
called the major axis.

Malus. A name applied to a portion of the constellation Argo.

Marfik. A name given by the Arabian astronomers
to the star λ Ophiuchi.

Markab. A name sometimes applied to the star
a Pegasi.

Mars. One of the primary planets. It revolves

round the sun at a mean distance of 141,000,000 miles in a period of 687 of our days. Its diameter is about 4200 miles, and markings on its surface are supposed to indicate the existence of land and water. It possesses an atmosphere, and may possibly be inhabited by some forms of life. It has two very small satellites, discovered by Professor Asaph Hall in August 1877. For further details, see Appendix.

Marsic. A name sometimes applied to the star κ Herculis.

Mass. The quantity of matter contained in a body. The weight of the same body would vary if placed on different planets and—very slightly—at different parts of the earth's surface, but the mass remains constant wherever the body is situated.

Mass of Binary Stars. The mass or quantity of matter contained in the components of a binary star is usually expressed in terms of the sun's mass taken as unity. If the distance of a binary star can be determined, we can find from its parallax and the elements of the computed orbit the mean distance between the components in terms of the sun's mean distance from the earth, and then, by an extension of Kepler's third law, the mass of the system in terms of the sun's mass may be found (or, more correctly, in terms of the combined mass of the sun and earth ; but the earth's mass being relatively very small it may be neglected). The method of calculation is very simple, and is as follows : Divide the computed semi-axis major by the parallax (both expressed in seconds of arc). The quotient will express the mean distance between the components in terms of the sun's mean distance from the earth. Now cube this quotient and divide the result by the square of the period expressed in years, and the result will be the combined mass of the components of the binary star in terms of the sun's mass.

Mass of Sun and Planet. The mass of a planet is usually expressed as a fraction of the sun's mass taken as unity. Sometimes the sun's mass is stated in terms of the earth's mass taken as unity.

Masym. A name sometimes applied to the star λ Herculis.

Maxima and Minima of Variable Stars. The *maximum* of a variable star is the brightest phase of its varying light, and the *minimum* the faintest.

Mazzaroth. An ancient name for the star Sirius.

Mean Distance. The mean or average distance of a body moving in an elliptic orbit from the focus of the ellipse in which the central body lies. It is therefore equal to the semi-axis major of the ellipse, which is a mean between the greatest and least distances of the revolving body from the focus.

Mean Motion. The velocity with which a moving body would describe a circular orbit having a radius equal to the " mean distance," and in the same period as in the real orbit. In the case of a binary star the mean angular motion is 360° divided by the period in years.

Mean Noon. The time of transit of the " mean sun " across the meridian.

Mean Solar Day. The interval of time which elapses between two successive " mean noons," or transits of the imaginary " mean sun " across the meridian.

Mean Solar Time is the hour angle of the " mean sun " converted into time at the rate of 1 hour to 15° or 4 minutes to 1°.

Mean Sun. The imaginary sun or point used in the regulation of "Mean Time," which see.

Mean Time. The time shown by ordinary clocks and watches. It is regulated by the motion of the "Mean Sun"—an imaginary sun, or rather point, which moves uniformly round the celestial equator.

Mebsuta. A name sometimes applied to the star ε Geminorum.

Medium, Resisting. See Ether.

Megrez. A name sometimes applied to the star δ Ursæ Majoris.

Mekbuda. A name sometimes applied to the star ζ Geminorum.

Menkab. A name sometimes applied to the star α Ceti.

Menkalinan. A name sometimes applied to the star β Aurigæ. From the Arabic *Menkib dhi-l'inân*.

Menstrual Equation. An apparent monthly displacement of the sun in longitude, due to the fact that the moon revolves round the centre of gravity of the earth and moon, and not round the earth's centre.

Merak. A name sometimes applied to the star β Ursæ Majoris.

Mercator's Projection. A projection of the sphere sometimes used in map drawing. It represents the sphere "as it might be seen by an eye carried successively over every part of it" (Sir John Herschel).

Mercury. The nearest of the planets to the sun. It revolves round the sun in a period of about 88 days, at a mean distance of about 36,000,000 miles. Its orbit has the greatest eccentricity (0·205) of all the primary planets. Its diameter is about 3000 miles. For further details, see Appendix.

5

Meridian, Celestial. The great circle of the celestial sphere which passes through the poles and the zenith of the place of observation.

Meridian, Prime. The meridian on the earth's surface from which longitudes east and west are reckoned. The meridian of Greenwich Observatory is usually taken as the prime meridian.

Meridian, Terrestrial. A " great circle " on the earth's surface which passes through the terrestrial poles.

Merope. One of the stars in the Pleiades; otherwise known as 23 Tauri.

Mesartim. A name sometimes applied to the star γ Arietis.

Meteoric Stones. Stones which occasionally fall from the sky. Of these there are many well-authenticated cases. They have been classed as follows: "Siderites are those in which iron predominates; Siderolites represent those which are formed of iron and stone in large degree; Ærolites are applied to those which are nearly all stone " (Denning).

Meteors. Luminous bodies which suddenly appear in the atmosphere and move with great rapidity. They are also known as " falling stars " or " shooting stars." They are usually very small bodies, which become incandescent by friction with the air, but are occasionally of considerable size, when they are called fireballs. The mean velocity of meteors is about thirty-four miles a second. The ordinary shooting stars usually become visible at a height of seventy or eighty miles, and disappear at a height of about fifty or fifty-five miles above the earth's surface. Fireballs, however, sometimes approach within five to ten miles.

Method of Least Squares. See LEAST SQUARES.

Metonic Cycle. A lunar cycle discovered by Meton and Euctemon, B.C. 432. They found that after a period of nineteen years "new moons" and "full moons" recurred on the same days of the year. The Metonic Cycle is 235 synodic months = 6939·69 days, or almost exactly nineteen tropical years.

Metre. A French unit of measure, originally fixed at one ten-millionth of the length of a quadrant of the earth's meridian. The length of the metre is 39·37079 English inches, or 3·28089 feet.

Micrometer. An instrument for measuring accurately small angles. There are various forms of micrometer, such as the filar micrometer, the parallel wire micrometer, the position micrometer, heliometer, etc.

Microscopes. Used for reading the graduation on the circles of astronomical instruments.

Microscopium (the Microscope). One of the southern constellations.

Midnight. The time of the sun's transit below the pole. This usually occurs when the sun is below the horizon.

Milky Way. The nebulous band or zone of light which encircles the heavens. It consists of myriads of small stars, probably mixed up with nebulous matter. It is also called the Galaxy.

Mimas. The inner satellite of Saturn, that nearest to the planet, round which it revolves in a period of 22 hrs. 37 mins., at a mean distance of about 117,000 miles. It was discovered by Sir William Herschel on Sept. 17th, 1789. Its diameter may be about 1000 miles, but this is uncertain. Its stellar magnitude is, according to Pickering, only 12·8, and as it is so near Saturn it can only be seen with telescopes of considerable power.

Minor Axis of Orbit. In an elliptic orbit the axis which passes through the centre of the ellipse at right angles to the major axis.

Mintaka. A name sometimes applied to the star δ Orionis. From the Arabic *mintakat al-djauzâ*, " the belt of the giant."

Mira. A name applied to the wonderful variable star o Ceti. At the maximum the magnitude varies from 1·7 to 5 at different maxima. At minimum, the star is 8½ or 9 mag. The mean period is about 331⅓ days, but is subject to irregularities. The star has a remarkable spectrum of the third type, in which bright lines have been seen by several observers.

Mirach. A name sometimes applied to the star β Andromedæ, and also ε Boötis.

Mirfak. A name sometimes applied to the star α Persei. From the Arabic *al-marfik*, " the elbow."

Mirzum. A name sometimes applied to the star β Canis Majoris.

Mizar. A name applied to the star ζ Ursæ Majoris ; sometimes also to ε Boötis.

Monoceros (the Unicorn). One of the constellations. It lies to the east of Orion, and the equator passes through it.

Month, Anomalistic. The period of revolution of the moon round the earth, with reference to the line of apsides of the lunar orbit. This period is 27 days 13 hrs. 18 mins. 37·4 secs.

Month, Nodical. The period which elapses between the passage of the moon through one of its nodes, and its passage through the same node again. This period is 27 days 5 hrs. 5 mins. 35·8 secs.

Month, Sidereal. The period of the moon's

rotation round the earth, with reference to the stars; its length is 27 days 7 hrs. 43 mins. 11·4 secs.

Month, Synodical. The period which elapses between two successive conjunctions of the moon with the sun. This period is 29 days 12 hrs. 44 mins. 2·7 secs.

Moon. The Earth's satellite. Its mean distance from the earth is 60·27 times the earth's equatorial radius, or about 238,854 miles; but owing to the eccentricity of the moon's orbit round the earth, this varies from 225,742 miles, when the moon is in *perigee*, to 251,968 miles when it is in *apogee*. The eccentricity of the orbit is about $\frac{1}{18}$, and its inclination to the plane of the ecliptic about 5° 8″. The moon's diameter is 2163 miles. Its mass is $\frac{1}{81}$ of the Earth's mass, and its density (that of water = 1) is 3·40.

Moon-culminating Stars. Stars lying near the moon's apparent path in the sky, used for determining the longitude of the place of observation by measurement of their angular distance from the moon's centre.

Moon's Parallactic Inequality. See INEQUALITY.

Mothallath, or **Rás-al-Mothallath.** An Arabic name for the star α Trianguli.

Motion, Accelerated. Motion in which the velocity is constantly increasing. A body falling to the earth is an example of accelerated motion.

Motion, Apparent and Real. The *Apparent* motion of a planet is its motion as seen from the earth. The *Real* motion is its actual motion in space round the sun. Thus, when a planet is " retrograding," its apparent motion is from east to west among the stars, whereas its real motion is from west to east.

Motion, Direct. The motion of a planet when it is moving from west to east among the stars.

Motion, Proper. The motion of a star on the celestial sphere, due to the real motion of the star in space. See PROPER MOTION.

Motion, Relative. The motion of one moving body relative to another. Thus, if two bodies are moving on parallel lines in the same direction, but with different velocities, their relative motion is the difference between the two velocities; but if in opposite directions, the relative motion is the sum of the velocities.

Motion, Retrograde. The motion of the planets in the sky when they *apparently* move from east to west among the stars, or contrary to that of their real motion, which is from west to east. This apparent retrograde motion is due to the earth's orbital motion round the sun combined with that of the planet. The term is also sometimes applied to the apparent diurnal motion of the sun, moon, planets, and stars from east to west, due to the earth's rotation on its axis.

Mountains, Lunar. In addition to the numerous "craters" visible on the moon's surface, there are also lofty ranges of mountains. Some of the most important of these are the Alps, the Caucasus, the Apennines, the Carpathians, the Pyrenees, the Rock Mountains (highest 25,000 feet), the Leibnitz Mountains (highest 26,000 feet, or more), etc. Compared to the moon's diameter, the lunar mountains are much higher than ours.

Mukdim. An Arabic name for the star ε Virginis, otherwise known as Vindemiatrix.

Muphrid. A name sometimes applied to the star η Boötis.

Mural Circle. An astronomical instrument used in observatories. It consists of a large, graduated circle, firmly fixed in the plane of the meridian. To this circle a telescope is attached, with which the observations are made. The circle is mounted on the face of a wall; hence its name.

Musca (the Fly). One of the southern constellations. It lies south of the Southern Cross.

N.

Nadir. The point in the celestial sphere vertically below an observer at any point on the earth's surface. Its direction is pointed to by the plumb line, and it corresponds to the *zenith* of a place at the antipodes. See ZENITH.

Nath. A name sometimes applied to the star β Tauri. Derived from the Arabic *al-nátih*, "the butting," referring to its position on the tip of the Bull's Horn.

Neap Tides. The tides which occur at the moon's "quadratures"—that is, at "first quarter" and "last quarter." The heights of "neap" and "spring tides" are about in the ratio of 4 to 10. See SPRING TIDES.

Nebulæ. The hazy spots of light visible in the sky with a telescope. They have been classed as follows : Annular nebulæ, elliptic nebulæ, spiral nebulæ, planetary nebulæ, and nebulous stars. To these may be added irregular nebulæ, like the great nebulæ in Orion and Argo. Clusters of stars are sometimes, but incorrectly, spoken of as nebulæ. True nebulæ are, for the most part, gaseous.

Nebular Hypothesis. A theory proposed by Laplace to explain the origin of the solar system. He supposed the sun and planetary system to have been formed by the cooling and condensation of a rotating nebulous mass, which originally extended beyond the orbit of Neptune. During the process of contraction he supposed that rings were detached from the parent mass, and that these rings afterwards consolidated into planets and satellites. Numerous arguments have been advanced for and against this hypothesis. A full discussion of the question will be found in the present writer's *Visible Universe.*

Nebulosity. Hazy light visible in telescopes which cannot be resolved into stars.

Nebulous Stars. Stars surrounded with nebulosity. These are very rare objects: ϵ and ι Orionis are examples.

Nekkar. A name sometimes applied to the star β Boötis.

Neptune. The outermost planet of the solar system (so far as is known at present). The telescopic discovery of Neptune was made by Galle at Berlin, on Sept. 23rd, 1846; but its probable existence had been previously predicted by Adams and Le Verrier, from a consideration of irregularities in the motion of Uranus. The mean distance of Neptune from the sun is about 2,789,000,000 miles, and its period of revolution $164\frac{3}{4}$ years. The diameter of Neptune is about 36,000 miles. It is not visible to the naked eye, its stellar magnitude being about the eighth; but it may be seen with any small telescope if its position is accurately known. Neptune is attended by one satellite, which revolves round its primary in 5 days 21 hrs. 2 mins., at a mean distance of about 262,000 miles. It is a very faint object, and can only be well seen in large telescopes; but it is probably of considerable size.

New Moon. When the moon is in conjunction with the sun, or has the same celestial longitude, it is called " new moon." The term is popularly applied to the phase of the moon when it first appears, as a thin crescent, to the east of the sun after sunset; but this application of the term is incorrect.

Newtonian Telescope. A form of reflecting telescope, in which the rays reflected from the large mirror are again reflected at right angles into the eyepiece by means of a small plane mirror.

Nihal, or **Al-nihál.** A name given by the Arabian astronomers to the star β Leporis.

Nodes. The points in which the orbit of a planet or comet cuts the plane of the ecliptic. The node at which the planet or comet is rising from the southern to the northern side of the ecliptic is called the *ascending node,* and that at which the moving body is passing from the northern to the southern side of the ecliptic, the *descending node.* The line joining these points is called *the line of nodes,* and is therefore the line of intersection of the two planes. In the orbits of binary stars, the " position angle " of the line of nodes can be found from the observations, but it is impossible to determine which of the nodes is the *ascending,* and which the *descending node.* In other words, we cannot say from the measures at which node the companion star is approaching the eye, and at which it is receding. It would be possible—at least, theoretically—to determine this by observation with the spectroscope.

Nodical Month. See Month, Nodical.

Nonagesimal Point. The altitude of the highest point of the ecliptic in the sky at any given instant.

Noon, Mean, and **Apparent.** Mean Noon is the time of transit of the " mean sun " across the meridian, and Apparent Noon, the time of transit of the apparent or true sun.

Norma (the Rule). One of the southern constellations.

Normal Disturbing Force. In the theory of perturbations, the component of the disturbing force, which acts along a normal to the curve—that is, at right angles to the tangent at the place of the body.

North Polar Distance. The angular distance of a celestial body from the north pole of the celestial sphere. It is equal to the complement of the declination, or 90 minus the declination.

Nova. A term applied to temporary stars, which see.

Nubecula Major. The larger of the two nebulous spots in the southern hemisphere, popularly known as the "Magellanic Clouds." It consists of a collection of small stars of various magnitudes, apparently associated with star clusters and nebulæ of various forms.

Nubecula Minor. The smaller of the two nebulous spots in the southern hemisphere, popularly known as the "Magellanic Clouds." It consists of a collection of small stars, star clusters, and nebulæ.

Nucleus. The central portion of the head of a comet. The term is also applied to the most condensed portion of the light of a nebula.

Number, Golden. See GOLDEN NUMBER.

Number of Eclipses. The number of eclipses of the sun and moon in the year may amount to seven, and must be at least two. If only two, they are both of the sun. If there are seven, five must be of the sun and two of the moon. There cannot be more than three eclipses of the moon in a year, and in some years there are none. In a period of eighteen years—the length of the Saros (which see)—there are usually about seventy eclipses, twenty-nine of the moon and forty-one of the sun.

Number of Fixed Stars. The number visible to very good eyesight does not much exceed 7000 for the whole sky. The total number visible in the largest telescopes does not probably exceed 100 millions. Owing to the possible extinction of light at great distances, or for some other reasons, the number of the *visible* stars is, and must necessarily be, strictly limited.

Nutation, Lunar. A variation or perturbation in the revolution of the celestial poles round the pole of the ecliptic, due to the action of the moon. Its period is the same as that of a sidereal revolution of the moon's nodes, or about 18 years 220 days.

Nutation, Monthly. A variation in the revolution of the celestial poles round the pole of the ecliptic, due to the changes in the moon's declination. Its period is half a month.

Nutation, Solar. A variation in the revolution of the celestial poles round the pole of the ecliptic, due to the changes in the sun's declination. Its period is half a tropical year.

O.

Oberon. The outermost satellite of Uranus, or that farthest from the planet. Its mean distance from the centre of the planet is about 389,000 miles, and its period of revolution 13 days 11 hrs. 7 mins. It can be well seen only with large telescopes, and its diameter is uncertain. Oberon was discovered by Sir W. Herschel on Jan. 11th, 1787.

Object Glass. The large glass of an astronomical telescope, or that nearest the object. It usually consists of two lenses either cemented together, merely touching each other, or—in large telescopes—separated by several inches.   The outer lens is of a double convex shape, and of crown glass; the inner is usually double concave, and made of flint glass.   In some binocular field-glasses the object glass consists of three lenses cemented together.

Objects, Test.   Celestial objects, such as faint stars, close double stars, etc., which form tests for the "light grasping" power and definition of telescopes.

Oblate Spheroid.   A solid formed by the rotation of an ellipse round its minor, or shorter, axis.   The figure of the earth is that of an oblate spheroid, its shorter axis being the axis of rotation.

Obliquity of the Ecliptic.   The angle between the plane of the equator and the plane of the ecliptic.   The present inclination is about $23°\ 27\frac{1}{2}'$, but it is subject to a cyclical change between the limits—according to Stockwell—of $21°\ 58'\ 36''$ and $24°\ 35'\ 58''$.   That the seasons are due to the obliquity of the ecliptic was taught by Diogenes of Apollonia, about 450 B.C.

Observatory.   A building erected for the purpose of observing the heavenly bodies.   The name is also applied to buildings constructed for making meteorological and magnetical observations.

Occultation.   When one celestial body passes in front of another so as to hide it from view of the observer, the body so hidden is said to be occulted.   The moon occasionally occults the planets, and, more frequently, the stars.   There are also occultations by the sun, but these are invisible owing to the intense brilliancy of that body.   Occultations of stars by the planets have also been observed, but these are of rare occurrence.

Octans (the Octant). One of the southern constellations. The south celestial pole is situated in this constellation.

Offing. The bounding line of the horizon as seen from any point on the earth's surface. The term is applied particularly to the sea horizon.

Okda. A name sometimes applied to the star a Piscium.

Opaque. A substance through which light cannot pass is said to be opaque.

Opera Glass. A small binocular telescope of low power used in theatres. A good opera glass is useful in astronomy for observing the brighter phases of variable stars.

Ophiuchus (the Serpent-Bearer). One of the constellations. It stretches from Hercules to Scorpio. The celestial equator passes through it.

Opposition. When the angular distance between the celestial bodies is 180° (measured on a great circle passing through the two bodies) they are said to be in *opposition.* At " full moon," the moon is in opposition to the sun, or, more correctly speaking, nearly in opposition, as it can be *exactly* in opposition only during the totality of a lunar eclipse.

Orbit. The imaginary curve in space which a body describes when revolving round another. Thus, the path of a planet or comet round the sun, of a satellite round its primary, or of one component of a binary star round the other (or round the centre of gravity of both), is called the *orbit* of the moving body.

Orientation. The direction of a map: a system of triangles, buildings, etc., with reference to the east point.

Orion (the Hunter). One of the finest of the

constellations As the celestial equator passes through the centre of the constellation it is visible from nearly all parts of the earth's surface.

Orionids. A meteor shower, visible about Oct. 18th to 20th in each year. The meteors seem to radiate from a point in Orion (90° + 15°). They are swift, with streaks.

Orrery. An instrument for representing the planets and their motions round the sun. Called after the Earl of Orrery.

Orthogonal Disturbing Force. In the theory of perturbations, the component of the disturbing force resolved at right angles to the plane in which the disturbed body is at the instant moving round the centre of force.

Orthographic Projection. A method of mapping the surface of a sphere, in which every point on the hemisphere is projected on its base by a perpendicular let fall on it. In this projection the central portions of the hemisphere are well shown, but those towards the base are much crowded and distorted.

P.

Pallas. One of the minor planets or asteroids revolving round the sun in orbits lying between those of Mars and Jupiter. It was discovered by Olbers on March 28th, 1802. It revolves round the sun in a period of 4·605 years at a mean distance of 2·768 times the earth's mean distance from the sun. When in opposition its magnitude is about the eighth. The orbit of Pallas is remarkable for its high inclination, which amounts to 34° 44'.

Parabola. One of the conic sections, which may be

supposed formed by a plane cutting a cone obliquely parallel to the side of the cone. The parabola is therefore not a closed curve, but has two branches which extend out to infinity. Some comets have been found to move in a parabola.

Parallactic Angle. See ANGLE OF SITUATION.

Parallactic Inequality of the Moon.
See INEQUALITY.

Parallactic Instrument. An old name for the equatorial telescope. See EQUATORIAL TELESCOPE.

Parallax. An *apparent* change in the position of a celestial object due to a *real* change in the observer's position.

Parallel. A small circle of a sphere. See SMALL CIRCLE.

Parallel Wire Micrometer. A form of micrometer having two parallel wires which can be made to approach each other by means of screws.

Parameter. Same as "Latus Rectum," which see.

Partial Eclipse. An eclipse of the sun or moon in which only a portion of the disc is hidden or darkened.

Pavo (the Peacock). One of the southern constellations. It lies between Octans and Telescopium.

Pegasus (the Winged Horse). One of the northern constellations. The so-called "Square of Pegasus" is formed by β, a, and γ Pegasi, and a Andromedæ.

Penumbra. In an eclipse of the moon, the partial shadow which borders the dark shadow of the earth. At the points of the moon's surface covered by the penumbra the sun is seen partially eclipsed. The lighter shade surrounding the darker portion or *umbra* of a sun-spot is also called the *penumbra*.

Periastron. The point in the *real* orbit of a binary

star at which the component stars are at their closest. This point does not always coincide with the point of nearest approach in the *apparent* orbit as seen from the earth. The periastron point may be found by joining the centre of the apparent ellipse with the principal star and producing it to meet the apparent ellipse. If this line is produced in the opposite direction it will meet the ellipse at the point of apoastron, which see.

Perigee. The point in the moon's orbit which is nearest to the earth.

Perihelion. The point in a planet's or comet's orbit at which it is nearest to the sun. This point lies at the extremity of the major axis.

Period, or Periodic Time. The time taken by a planet or comet to revolve round the sun, or by a satellite to revolve round its primary. The term is also applied to the time in which the components of a binary star revolve round their common centre of gravity, and also to the time which elapses between two maxima and two minima of a variable star.

Period, Julian. See JULIAN PERIOD.

Periodical Stars. Another name for Variable Stars, which see.

Perpetual Day. The period in the Arctic zone when the sun does not set.

Perpetual Night. The period in the Arctic regions when the sun does not rise.

Perseids. A meteoric shower, which seems to radiate from the constellation Perseus (44° + 56°). The meteors appear about August 9th to 11th in each year. They are swift, and leave streaks.

Perseus. One of the northern constellations.

Personal Equation. The error in the observations

of the time of transit of a celestial body by a particular observer is called his "personal equation." The term might also be applied to other observations, such as the relative brightnesses of white and coloured stars, etc.

Perturbations. Inequalities produced in the orbital motion of the moon, planets, satellites, and comets by the attraction of the sun and the mutual attraction of each other.

Phact. A name sometimes applied to the star a Columbæ.

Phase. The particular aspect of a celestial body, the appearance of which is subject to periodical changes. Thus, we speak of the phases of the moon, the phases of the inferior planets, the phase of a lunar eclipse, etc.

Phecda. A name sometimes applied to the star γ Ursæ Majoris.

Phobos. The inner satellite of Mars. It revolves round the planet in about 7 hrs. 39 mins. Its distance from the centre of Mars is about 5,819 miles, and its diameter probably not more than 7 miles. It was discovered by Professor Asaph Hall on August 17th, 1877. For further details, see Appendix.

Phœnix (the Phœnix). One of the southern constellations.

Photography, Stellar. The art of photography applied to the mapping of the stars. Owing to the introduction of the "dry plate process," and the manufacture of very sensitive plates, it is now possible to obtain photographs of stars, nebulæ, etc. Most beautiful photographs of stars, star clusters, and nebulæ have been obtained by the brothers Henry at the Paris Observatory, and by Dr. Common and Dr. Roberts in England. Photographs are now being taken

6

at several observatories on an organised plan for the construction of a photographic chart of the whole heavens.

Photometer. An instrument for measuring the relative brightness of the stars. These are of various forms, but those now most generally used are the " wedge photometer," used at the Oxford Observatory, and the so-called " meridian photometer," used at the Harvard Observatory (U.S.A.). Catalogues of the magnitudes of the brighter stars, as measured with the photometer, have been published by these observatories. That constructed at Oxford is called the *Uranometria Nova Oxoniensis*, and that found at Harvard the *Harvard Photometry*.

Photometric Scale. The scale in which the brightness of the stars is represented according to a fixed standard. The number which expresses the number of times which the light of a given star exceeds that of another one magnitude fainter is termed the "light ratio." The number now universally adopted by astronomers is 2·5119, of which the logarithm is 0·4.

Photometry of the Stars. The measurement of the relative brightness of the stars, by means of instruments specially designed for the purpose. See PHOTO-METER.

Phurud. A name sometimes applied to the star ζ Canis Majoris.

Pictor (the Painter's Easel). One of the southern constellations.

Pisces (the Fishes). One of the zodiacal constellations. Owing to the precession of the equinoxes, the " First Point of Aries" now lies in the constellation Pisces.

Piscis Australis (the Southern Fish). One of the southern constellations. Its brightest star is Fomalhaut.

Places, Geocentric and Heliocentric. The geocentric place of a celestial body is its position on the star sphere as supposed to be seen from the centre of the earth, and its heliocentric place its position as seen from the centre of the sun.

Places, Star. The correct position of the stars on the surface of the celestial sphere. These are usually defined by stating their right ascensions and declinations for a given epoch. See RIGHT ASCENSION and DECLINATION.

Planetary Motion. The motion of the planets from west to east, or contrary to the diurnal motion. The reality of this motion was taught by Alemæon of Croton, in the fifth century B.C.

Planetary Nebulæ. Nebulæ of a uniform, or nearly uniform, brightness—at least when viewed with telescopes of moderate power—and usually of a circular or elliptical shape, with discs resembling those of the planets, but of course very much fainter. Some are of a pale blue colour.

Planets, Minor, or **Asteroids.** The group of small planets 'which revolve round the sun in orbits lying between those of Mars and Jupiter. They are very small bodies. The diameter of the largest, Vesta, probably does not exceed 200 miles. The number now known (1893) amounts to over 300.

Planets, Primary. The planets which revolve round the sun as a centre. These are in order of distance from the sun: (1) Mercury, (2) Venus, (3) the Earth, (4) Mars, (5) the Group of Minor Planets, (6) Jupiter, (7) Saturn, (8) Uranus, (9) Neptune.

Planets, Secondary. The satellites which revolve

round the primary planets as a centre. Our moon is a secondary planet, or satellite of the earth, but from its relatively large size and other reasons, it may be almost considered as a primary planet. Mars has 2 satellites, Jupiter 5, Saturn 8, Uranus 4, and Neptune 2 : a total of 22 secondary planets.

Platonic Period. The period of revolution of the equinoxes : about 25,695 years.

Pleiades. The well-known group or naked-eye cluster of stars surrounding the third-magnitude star Alcyone or η Tauri. To ordinary vision, perhaps, only six stars can be seen distinctly with the naked eye ; but to keener eyesight more are visible. With a good opera-glass over thirty may be seen. Powerful telescopes show several hundred, and on a photograph taken at the Paris Observatory no less than 2326 may be counted. A quantity of nebulous light is also visible on the photograph, surrounding the brighter stars of the group.

Plumb-line. A weight suspended by a cord. It hangs exactly perpendicular to the surface of smooth water, and therefore perpendicular to a tangent to the earth's surface at the place of observation.

Pointers. A term applied to the stars a and β Ursæ Majoris (or " the Plough ") because they nearly point to the Pole Star.

Points of Compass. The principal points of the compass are North, South, East, and West. These are called the Cardinal Points. Each quadrant of 90° is, however, further subdivided into 8 divisions, or $11\frac{1}{4}°$ to each division, making 32 points in all. These are designated as follows, beginning at the north, and going round the circle by east, south, and west, back to north again : N., N. by E., N.N.E., N.E. by N., N.E., N.E. by

E., E.N.E., E. by N., E., E. by S., E.S.E., S.E. by E., S.E., S.E. by S., S.S.E., S. by E., S., S. by W., S.S.W., S.W. by S., S.W., S.W. by W., W.S.W., W. by S., W., W. by N, W.N.W., N.W. by W., N.W., N.W. by N., N.N.W., N. by W., N.

Polar Distance. The angular distance of a celestial body from one of the poles of the celestial sphere. The distance from the north pole is called the north polar distance, and that from the south pole the south polar distance.

Polaris, or **Pole Star.** The nearest bright star at present to the north celestial pole. It is otherwise known as α Ursæ Minoris. Its present distance from the pole is about $1\frac{1}{2}°$, and the distance is diminishing.

Poles, Celestial. The poles of the celestial sphere are the points towards which the earth's axis of rotation points. They are, in fact, the extremities of an imaginary axis round which the star sphere *apparently* rotates.

Pole Star, or **Polaris,** which see.

Poles, Terrestrial. The extremities of the earth's axis of rotation, or the points at which the axis meets the surface.

Pollux. A name applied to the bright star β Geminorum.

Pores of Sun's Surface. The minute dark spots visible on the sun's surface with a telescope.

Porrima. A name sometimes applied to the star γ Virginis.

Position Angle. The angle between the line joining the components of a double star, and the "declination circle" passing through the primary star of the pair. This angle is measured from 0° to 360°, beginning at the north point (or bottom of the field in an inverting telescope) and going round by east, south, and west. In a binary star the motion is said to be *direct* when the position angle is increasing numerically, and retrograde when diminishing.

Position Micrometer. A form of parallel-wire micrometer used in measuring double stars.

Postvarta. A name sometimes applied to the star γ Virginis.

Præsepe (the Bee-hive). The star cluster or group of small stars in the constellation Cancer (the Crab).

Precession of the Equinoxes. A slow change in the position of the celestial equator, which causes the equinoctial points to retrograde along the ecliptic. This is due to the pole of the equator revolving round the pole of the ecliptic in a period of about 25,695 years (Stockwell). This motion of the plane of the equator is due to the disturbing effect of the attractions of the sun and moon on the protuberant matter at the earth's equator. It was discovered by Hipparchus in the second century B.C.

Primary Planets. The planets which revolve round the sun as a centre. See PLANETS, PRIMARY.

Prime Meridian. The meridian on the earth's surface from which longitudes east and west are reckoned. The meridian of Greenwich Observatory is usually taken as the prime meridian, but the French reckon from the meridian of Paris.

Prime Vertical. The "great circle" of the celestial sphere which passes through the zenith, nadir, and west points of the horizon.

Priming of the Tides. An acceleration in the time of high water which occurs between "new moon" and "first quarter," and between "full moon" and "last quarter," due to the combined action of the sun and moon.

Problem of Three Bodies. When a large central body has two smaller bodies revolving round it, the investigation of the perturbations of the system thus formed is called the "problem of three bodies." Its *exact* solution is beyond the present powers of mathematical analysis; but when the central body is very large in comparison with the others, as in the case of the sun and planets, a sufficiently close approximation can be made.

Procyon. A name applied to the bright star α Canis Minoris. Derived from the Greek προκύων, "the advanced dog," or the dog which goes before Sirius "the great dog."

Projections of the Sphere. Methods of mapping the surface of a sphere on a plane (or flat) surface.

Prolate Spheroid. A solid formed by the rotation of an ellipse round its major or longer axis.

Proper Motions. Many of the so-called "fixed stars" are not really fixed, but have a small motion across the face of the sky. This is called the star's "proper motion." The motion is in many cases very small, but becomes perceptible with accurate astronomical instruments, after the lapse of a number of years. Proper motion is due to a real motion of the star, combined with an apparent motion due to the sun's motion through space.

Pulcherrima. A name sometimes applied to the beautiful double star ε Boötis.

Puppis. A name applied to a portion of the constellation Argo.

Q.

Quadrant. The fourth part of a circle, or a quarter circle.

Quadrantids. A shower of meteors visible about Jan. 2nd in each year. They seem to radiate from a point north of Corona Borealis. They are swift, with long paths.

Quadrature. A term applied to the position of two celestial bodies when the difference of their longitudes is 90°. The moon is in quadrature at " first quarter " and " last quarter."

Quarter, First, and Last. Terms applied to the phases of the moon when the disc is half illuminated. This occurs when the moon's angular distance from the sun is 90°. These phases are also called " half moon."

R.

Radial Disturbing Force. In the theory of perturbations, the component of the disturbing force which acts along the radius vector.

Radiant. The point in the celestial sphere from which a shower of meteors seems to radiate.

Radiation, Solar. The amount of heat received from the sun by any particular planet.

Radius Vector. A line supposed to be drawn from a moving body to the centre round which it moves. In a circular orbit the radius vector is constant and equal to the radius of the circle, but in an elliptic orbit it varies in length with the position of the moving body.

Rasalas. A name sometimes applied to the star μ Leonis. From the Arabic *Rás-al-Asad*.

Ras Algethi. A name sometimes applied to the star α Ophiuchi.

Ras Alhague. A name sometimes applied to the star α Ophiuchi.

Rate of Clock. The amount by which an astronomical clock gains or loses in twenty-four hours. If the clock loses the rate is positive; if it gains it is negative.

Reading Microscopes. Small microscopes used for reading the graduated circles of astronomical and other instruments.

Real Ellipse. The actual ellipse in space described by one of the components of a double star round the other, supposed to be at rest. The orthogonal projection of this ellipse on the background of the sky is the "apparent ellipse," as seen from the earth. The real ellipse is only seen by a terrestrial observer when the plane of the orbit lies at right angles to the line of sight, and such cases are very rare.

Red Stars. Stars of a very reddish colour. For list of the most remarkable red stars, see Appendix.

Reflecting Circle. An instrument invented in 1770 by Tobias Mayers (Prof. Johann T. Mayer), and afterwards improved by the Chevalier de Borda in France and Mr. Edward Troughton in England. The principle of its construction is that of the sextant, but the graduated arc is a complete circle.

Reflecting Telescope. A form of telescope in which the image is formed by reflection from a concave mirror, and again reflected into the eyepiece by means of a smaller mirror. There are four forms of reflecting telescope : viz., the Newtonian, the Gregorian, the Cassegrainian, and the Herschelian, or "front view," which see.

Reformation of Calendar. An improvement in the method of reckoning time, first introduced by Julius Cæsar, B.C. 44, and in later times further corrected by

Pope Gregory XIII. See CALENDAR, GREGORIAN, and JULIAN.

Refraction. The bending or change of direction which a ray of light suffers when passing through a transparent medium, like the earth's atmosphere, or the object-glass of a refracting telescope.

Refracting Telescope. A form of telescope in which an image formed by refraction through the object glass is viewed by an eyepiece placed at the other end of the tube. The largest refracting telescope yet made (1893) is that at the Lick Observatory, California. The object-glass is 36 inches in diameter.

Regulus. A name applied to the bright star a Leonis.

Repetition. A method of measuring an angle on a graduated circle, invented by Borda. The method consists in repeating the measure several times along the graduation, and dividing the final reading by the number of observations. Thus, supposing the angle to be measured is approximately say 15°, then we measure from zero to 15°, then from 15° to about 30°, from 30° to 45°, and so on continuously without changing the index. Suppose the final reading to be 121° 20' and the number of observations 8, then the correct angle will be

$$\frac{121° \ 20'}{8} = 15° \ 10'.$$

Reticulated Micrometer. A form of micrometer having a series of wires crossing each other at right angles.

Reticulum (the Net). One of the southern constellations.

Retrograde Motion. See MOTION, RETROGRADE.

Reversal. A method of testing the adjustment of

the collimation in a transit instrument by reversing the telescope in its supports, so that the eastern end of the axis shall lie in the western support, and *vice versâ*.

Revolution. The motion of one body round another, or round the common centre of gravity of both bodies. Revolution should be carefully distinguished from *Rotation*, which means the motion of a body round a fixed axis contained within the body itself.

Rhea. One of the satellites of Saturn, the fifth in order counting from the planet, round which it revolves at a mean distance of about 336,000 miles, in a period of 4 days 12 hrs. 25 mins. Its diameter is somewhat doubtful, but its stellar magnitude is, according to Professor Pickering, 10·8. Rhea was discovered by J. D. Cassini on Dec. 23rd, 1672.

Rigel. A name applied to the bright star β Orionis. Derived from the Arabic *Ridjl-al-djauzâ*, "the giant's leg."

Right Ascension. The angular distance of a fixed star or other celestial body measured from the "First Point of Aries" eastward *on the equator*. This, combined with the declination, which is measured north and south from the equator, on a great circle passing through the celestial poles and the body, fixes the position of the body on the star sphere.

Rings of Saturn. A marvellous system of flat rings surrounding the planet Saturn, poised in space, and nowhere touching the planet. Various theories of their constitution have been advanced, but the most probable one is that they consist of a multitude of small satellites, too small to be individually visible, even with the most powerful telescopes. The rings are comparatively very thin, possibly not more than fifty miles in thickness·

For dimensions of the ring system, see Appendix ; and for fuller details, see popular works on Astronomy.

Rising of Celestial Objects. The appearance of a celestial body above the horizon of the place of observation. The time of rising is accelerated by refraction, which causes the object to appear above the horizon when it is actually below it. See REFRACTION.

Rotanev. A name sometimes applied to the star β Dolphini. Webb supposes it to be the name " Venator " reversed.

Rotation. The motion of a body round a fixed axis contained within the body itself. Rotation should be carefully distinguished from *revolution*, which means the motion of one body round another or round the common centre of gravity of both bodies.

S.

Sadachbia. A name sometimes applied to the star γ Aquarii.

Sadalmelik. A name sometimes applied to the star a Aquarii. From the Arabic *sad-al-malik*, " the good fortune of the king" (!).

Sadalsund. A name sometimes applied to the star β Aquarii. From the Arabic *sad-al-suûd*, " the fortune of fortunes "; a term given to the stars β and ξ Aquarii by the old Arabian astronomers.

Sagitta (the Arrow). One of the northern constellations. It lies between Vulpecula and Aquila.

Sagittarius (the Archer). One of the zodiacal constellations.

Saros. A lunar cycle discovered by the Chaldæan astronomers. It is the period of revolution of the nodes of the moon's orbit with reference to the sun, called the synodic revolution of the nodes. This period is 346·644 days. Now, nineteen synodic revolutions = 6,586·236 days, and is nearly equal to 223 lunar months, which amount to 6,585·29 days, or 18 years and 11 days. Hence, in this period the solar eclipses will be nearly the same. During the period of the Saros the total number of eclipses is about seventy—twenty-nine of the moon and forty-one of the sun.

Satellites. The smaller bodies which revolve round the planets of the solar system. The moon is a satellite of the Earth. Mars has 2 satellites, Jupiter 5, Saturn 8, Uranus 4, and Neptune 2—a total of 22.

Saturn. Next to Jupiter, the largest planet of the solar system. It revolves round the sun in a period of 29 years 167 days, at a mean distance of about 885,000,000 miles. Its mean diameter is about 72,000 miles, or about nine times that of the Earth. It therefore exceeds the Earth in volume over 700 times; but in density it is very light, its mass being only 94 times the mass of the Earth. It is surrounded by a wonderful system of thin rings, which forms the most unique and interesting phenomenon in the solar system. For further details see Appendix.

Scheat. A name sometimes applied to the star β Pegasi.

Schedir. A name sometimes applied to the star α Cassiopeiæ. Probably a corruption of the Arabic *al-sadr*, "the beast." The star is slightly variable in light.

Scintillation. A term sometimes applied to the twinkling of the stars.

Scorpio (the Scorpion). One of the southern zodiacal constellations.

Sculptor (the Sculptor's Workshop). One of the southern constellations.

Seasons. The variation in the relative length of the day and night, due to the inclination of the earth's axis of rotation to the plane of its orbit round the sun. That the seasons are caused by the inclination of the earth's axis was taught by Diogenes of Apollonia about 450 B.C.

Secondary. A term applied to the satellites which revolve round the planets of the solar system. The term is also applied to the great circles on a sphere which pass through the poles of another circle.

Sections, Conic. See CONIC SECTIONS.

Sector, Dip. See DIP SECTOR.

Sector, Zenith. An instrument for measuring the zenith distance of stars. Invented by Hooke in 1669.

Secular Acceleration of the Moon's Mean Motion. An acceleration in the moon's motion, or shortening in its period of revolution round the earth. The moon's mean motion increases at the rate of about eleven seconds in a century. The acceleration is due partly to the variation in the eccentricity of the. earth's orbit, and partly to a slight increase in the length of the sidereal day.

Secular Variations. Inequalities in the motions of the planets which do not depend on the configurations of the planets with reference to each other. The effect of secular variations is only perceptible after long periods of time. One of the most important of the secular variations is the slow increase and decrease in the eccentricity of the earth's orbit.

Secunda Giedi. A name sometimes applied to the star a^2 Capricorni. .

Selenography. The study of the moon's surface.

Serpens (the Serpent). One of the constellations.

Sexagesimal. The division of the circumference of a circle into 360 degrees. Each degree is subdivided into 60 minutes, and each minute into 60 seconds.

Sextans (the Sextant). One of the constellations. It lies between Leo and Hydra.

Sextant. A mathematical instrument used for measuring angles. It consists of a graduated arc of a circle, fitted with two mirrors and a small telescope. The principle of the instrument depends upon the following optical property:—"The angle between the first and last directions of a ray which has suffered two reflections in one plane is equal to twice the inclination of the reflecting surfaces to each other." The instrument is chiefly used in navigation, but also occasionally for astronomical purposes.

Shadow. The shade cast by an opaque body.

Sháulah. A name sometimes applied to the star λ Scorpii. Derived from the Arabic *al-schaulat*, and *schaulat al-akrab*, "the tail of the scorpion." The stars λ and υ Scorpii were also called *al-ibrat*, "the sting."

Sheliak. A name sometimes applied to the star β Lyræ.

Sheratan. A name sometimes applied to the star β Arietis.

Shooting Stars. A term applied to meteors or "falling stars," which are occasionally seen to shoot across the sky. "Shooting stars" are usually very small bodies, and have no connection with the fixed stars.

Sidereal. Relating to the stars.

Sidereal Month. The period of the moon's revolution round the earth, with reference to the stars. Its length is 27 days 7 hrs. 43 mins, 11·4 secs.

Sidereal Noon. The time of transit of the " First Point of Aries" across the meridian.

Sidereal Period. The period of revolution of a planet round the sun, with reference to the stars.

Sidereal Time. The time measured by the apparent rotation of the star sphere, or transit of the " First Point of Aries " across the meridian.

Sidereal Year. The time which elapses between two successive returns of the sun to the same position among the fixed stars. The length of the sidereal year is 365 days 6 hrs. 9 mins. 8·97 secs. It is therefore about twenty minutes longer than the tropical year, and about 4½ minutes shorter than the anomalistic.

Siderites. Meteoric stones which are chiefly composed of iron.

Siderolites. Meteoric stones containing a mixture of iron and stone.

Signs of Zodiac. The twelve constellations through which the ecliptic passes. These are :—1. Aries (the Ram); 2. Taurus (the Bull); 3. Gemini (the Twins); 4. Cancer (the Crab); 5. Leo (the Lion); 6. Virgo (the Virgin); 7. Libra (the Balance); 8. Scorpio (the Scorpion); 9. Sagittarius (the Archer); 10. Capricornus (the Goat); 11. Aquarius (the Water-bearer); and 12. Pisces (the Fishes).

Sirius. The star α Canis Majoris, the " dog star." It is the brightest star in the sky, being about two magnitudes brighter than an average star of the first magnitude, like Altair or Spica.

Sirrah. A name formerly applied to the star α Andromedæ.

Situation, Angle. The angle between the circles of declination and of latitude passing through a given star.

Skat. A name sometimes applied to the star δ Aquarii.

Small Circle. A circle on a sphere of which the plane does not pass through the centre of the sphere. Small circles are also called *parallels*.

Solar. Relating to the sun.

Solar Cycle. A cycle consisting of twenty-eight Julian years, at the end of which period the days of the week return to the same days of each month throughout the year.

Solar Day. The interval of time between two successive noons, or two successive midnights.

Solar System. The system of planets with their satellites, and comets, which revolve round the sun as a centre.

Solar Time. The time measured by the passage of the sun across the meridian. The time of transit is called *Apparent Noon*.

Solar Year. Same as Tropical Year, which see.

Solstice. Points on the ecliptic which lie at the maximum distance north and south of the celestial equator. The northern point is called the summer solstice, and the southern the winter solstice. The summer solstice is situated in Gemini, and the southern solstice in Sagittarius.

Southing. A term applied to the transit of a celestial body across the meridian of the place of observation, and to the south of the zenith.

South Polar Distance. The angular distance of a

celestial body from the south pole of the celestial sphere.

Specific Gravity. The ratio of the weight of a substance to that of an equal volume of water. Thus the earth's specific gravity is about $5\frac{1}{2}$, which implies that its weight is five and a half times that of a globe of water of the same size.

Spectrum Analysis. " The determination of the constituent elements of a luminous body by the examination of its light after its passage through one or more prisms " (Chambers' *Descriptive Astronomy*). Dark lines are visible in the spectra of the sun and stars; and from a comparison of these lines with the bright lines in the spectra of incandescent terrestrial substances, it is possible to determine the chemical elements present in the sun and stars. The *dark* lines in the solar spectrum and in stellar spectra are due to the absorption of the light of the incandescent element when shining through its own vapour. Spectrum analysis is now much used in astronomical researches for the purpose of dividing the stars into classes, and also for determining their motion in the line of sight. This research is much aided by photography.

Speculum. A term applied to the large mirror of a reflecting telescope. The speculum may be formed either of polished metal, or of a glass disc ground to the proper curve, and then silvered over and polished. The latter form of speculum is the one now generally used, and is called " silver on glass."

Sphere. A solid which may be supposed formed by the rotation of a circle round one of its diameter. Every point on the surface of a sphere is equidistant from the centre.

Spheroid. A solid formed by the rotation of an ellipse round one of its axes. If the rotation takes place round the minor axis, the solid is called an *oblate spheroid;* if round the major (or longer) axis, a *prolate spheroid.*

Spica. A named applied to the bright star a Virginis.

Spring Tides. The high tides which occur at new and full moon. The heights of spring and neap tides are about in the ratio of 10 to 4.

Spots on Sun. See Sun Spots.

Stars. The brilliant points of light visible in the sky at night. The stars are of all degrees of brilliancy, from Sirius, the brightest star in the heavens, down to the faintest point visible in the largest telescopes on the clearest nights. They have been divided into magnitudes, the first magnitude including the brightest stars, the second those decidedly fainter, and so on, down to the seventeenth magnitude, which is, perhaps, the faintest which has yet been *seen* with any telescope. Possibly, however, fainter stars have been photographed. Each magnitude is further subdivided decimally. Thus we have stars of magnitude 2·1, 2·2, 2·3, etc.

Stars, Binary. See Binary Stars.

Stars, Double. Stars so close that they appear as single stars to the naked eye. Some of these are so excessively close that it requires the largest telescopes to divide them. On the other hand, some may be seen with an opera glass, or even with the naked eye; but these cannot properly be called double stars. Some real double stars—that is, binary or revolving double stars—may, however, be seen with small telescopes, when the components are at their greatest distance apart. Of these, a Centauri and γ Virginis are examples.

Stars, Variable. See VARIABLE STARS.

Stationary Points. The points in a planet's orbit at which the planet appears stationary among the stars as seen from the earth. In some books on astronomy it is stated that a planet is stationary when it is moving directly towards or away from the earth; but this is quite incorrect.

Stereograms. Photographic views of the moon taken at different phases of libration. These combined in a stereoscope give the effect of a spherical body.

Stereographic Projection. A method of mapping the surface of a sphere, in which the eye is supposed placed at the extremity of a diameter of the sphere, and objects on the opposite hemisphere are projected on a plane passing through the centre of the sphere, and at right angles to the diameter passing through the eye.

Stones, Meteoric. See METEORIC STONES.

Style, Old and New. Prior to 1582 the year commenced on March 25th. The new style, in which the year commences on Jan. 1st, was not introduced into Great Britain till 1752, and as 170 years had elapsed since the new style was established by Pope Gregory, it was necessary to get rid of 11 days; and this was done by calling Sept. 3rd Sept. 14th. In Russia the old style is still retained.

Sub-Polo. A term applied to the passage of a celestial body across the meridian below the pole—that is, between the pole and the north point of the horizon.

Sub-solar Point. The point on the earth's surface at which the sun is in the observer's zenith on any given day at any given moment of Greenwich time.

Suhà. A name applied by the ancients to the star Alcor, near Mizar (ζ Ursæ Majoris).

Sulaphat. A name sometimes applied to the star γ Lyræ.

Summer Solstice. The point on the ecliptic at which the sun attains its maximum distance north of the celestial equator. This point is reached by the sun on June 21st, which is popularly known as the " longest day."

Sun. The centre of the planetary system. Its mean distance from the earth is about 92,796,950 miles (Harkness), and its diameter about 866,000 miles. Its density or specific gravity is 1·40 (that of the earth being 5·6, and water equal to 1). The sun revolves on its axis in a period of about $25\frac{1}{4}$ days. The axis is inclined to the plane of the ecliptic at an angle of about 83°, and points nearly to the fifth-magnitude star π Draconis. For further particulars see Appendix.

Sun-dial. An instrument for showing the time by means of a shadow cast by the sun on a dial plate. The rod or plate which casts the shadow is called the *gnomon* or *style*. It is placed parallel to the Earth's axis, and therefore points to the celestial pole. There are several forms of sun-dial. The sun-dial is a very ancient form of time-piece, and is mentioned in the Bible (with reference to the sickness of Hezekiah). A sun-dial was erected by Anaximander at Sparta, B.C. 545; one by Meton at Athens, B.C. 433; and one at Rome by Papirius Cursor, B.C. 306. The sun-dial shows apparent time, and to obtain mean time the time indicated by the sun-dial must be corrected by the Equation of Time, which see.

Sun Spots. Dark spots visible at times on the sun's surface. They usually consist of a dark central portion, called the *umbra*, surrounded by a lighter shade, called the *penumbra*. Several instances of spots large enough to be visible to the naked eye have been recorded. The display of sun spots is subject to a periodical variation, the maxima occurring at intervals of about eleven years, with intermediate minima. A maximum occurred in 1882, and another in 1893.

Superior Conjunction. When Mercury and Venus are in that part of their orbit beyond the Sun, as seen from the Earth, they are said to be in superior conjunction.

Superior Planets. The planets of the solar system which are farther from the Sun than the Earth. With the exception of Mercury and Venus, all the planets are superior.

Svalocin. A name sometimes applied to the star *a* Delphini. Webb supposed it to be the name "Nicolaus" spelt backwards.

Sweeps. A term employed by Sir William Herschel to denote his observations of the number of stars visible in various parts of the sky when the telescope was clamped, and the stars were allowed to pass through the field of view by the effects of the diurnal motion.

Synodical Month. The period which elapses between two successive conjunctions of the moon with the sun. Same as a *lunation*. Its length is 29 days 12 hrs. 44 mins. 2·7 secs.

Synodic Period. The period which elapses between two successive conjunctions or oppositions of a planet with the Sun.

Synodic Revolution. Same as Synodic Period, which see.

Synodic Rotation of the Sun. The *apparent*

period of the sun's rotation on its axis. Owing to the revolution of the earth in its orbit in the same direction as the sun's rotation, the apparent period is about two days longer than the real period. The synodical period of rotation is about 27 days 6 hrs. 40 mins.

Synodic Year. Twelve lunar months, or about 355 days. It is a term not often used.

System. Two or more celestial bodies revolving according to the laws of gravitation are said to form a "system." Thus, we have the solar system, Saturn's system, binary star systems, etc.

Syzygy. The moon is said to be in syzygy when it is in conjunction with the sun or "new moon," or when it is in opposition to the sun at "full moon."

T.

Talita. A name sometimes applied to the star ι Ursæ Majoris.

Tangential Force. In the theory of perturbations the component of the disturbing force which acts along the tangent to the orbit of the disturbed body, drawn in the plane of the orbit at the place of the body.

Tangent Screw. A screw used for giving a slow motion to a graduated arc after it has been clamped to the vernier. It was invented by Helvetius about the year 1650. See VERNIER.

Tarazed. Another name for γ Aquilæ. Derived from the Arabic *shâhin târâzed*, "the soaring falcon."

Taurids. A meteor shower visible about November 1st to 8th in each year. The meteors seem to radiate from a point in Taurus (58° + 20°). They are slow and brilliant. Another shower from the same region (62° + 22°) appears about November 20th and 27th.

Taurus (the Bull). One of the zodiacal constellations. Its brightest star is Aldebaran (a Tauri). The Pleiades and Hyades are in this constellation.

Taygeta. One of the stars (19 Tauri) in the Pleiades.

Tegmine. A name sometimes applied to the star ζ Cancri.

Tejat Post. A name sometimes applied to the star μ Geminorum. It comes from the Arabic *tahyáh*.

Telescope. An astronomical instrument for observing the heavenly bodies. It magnifies the image of the object observed, and thus brings it apparently nearer the eye. There are two forms of telescope—the Refracting Telescope and the Reflecting Telescope, which see.

Telescopic Objects. Celestial objects which cannot be seen with the naked eye, but require a telescope or opera-glass to render them visible.

Telescopium (the Telescope). One of the southern constellations.

Temporary Stars. Stars which blaze out suddenly, and after remaining visible for a short time fade away and become very faint stars, or planetary nebulæ, or totally disappear. They are also called *novæ*. Temporary stars are exceedingly rare objects. The recorded instances in modern times are those of 1572 in Cassiopeia (Tycho Brahé's "Pilgrim Star"); 1604 in Ophiuchus ("Kepler's nova"); 1670 in Cygnus (Anthelm's); 1848 in Ophiuchus (Hind's); 1866 in Corona Borealis (Birmingham's); 1876 in Cygnus Schmidt's); 1885 in the great nebula in Andromeda; and 1892 in Auriga (Anderson's). All these, with the exception of the "Blaze Star" of 1866, appeared in or near the Milky Way.

Terminator. A term applied to the line—usually

irregular—which divides the bright or illuminated part
of the moon from the dark part. At new and full moon
the terminator coincides with the limb or circumference
of the disc.

Terrestrial Equator. The great circle on the
earth's surface, every point on which is equidistant from
either pole. The plane of the equator is at right angles
to the earth's axis of rotation.

Terrestrial Latitude. The angular distance of a
place on the earth's surface north or south of the terrestrial equator. This is measured from 0° to 90°. Thus,
the latitude of the equator is 0°, and that of the poles
90°. The altitude of the celestial pole is equal to the
latitude of the place of observation.

Terrestrial Longitude. The angular distance of
a place on the earth's surface east or west of a fixed
meridian called the first or prime meridian. The meridian
of Greenwich is usually taken as the prime meridian.
Longitudes are measured east and west from 0° to 180°.

Terrestrial Meridian. The meridian of any place
on the earth's surface is the great circle passing through
the place, and the terrestrial poles. It therefore passes
through the earth's axis.

Terrestrial Poles. The extremities of the earth's
axis of rotation, or the points at which this axis meets
the surface. That situated in the hemisphere containing
Europe is called the north pole and the opposite the
south pole.

Tethys. One of the satellites of Saturn, the third
in order counting from the planet, round which it
revolves in a period of 1 day 21 hrs. 18 mins., at a mean
distance of about 187,000 miles. It was discovered by
J. D. Cassini in March 1684. Its stellar magnitude is,

according to Pickering, 11·4, but its real diameter is uncertain.

Thuban. A name sometimes applied to the star α Draconis.

Tides. The daily rise and fall of the waters of the ocean caused by the attraction of the sun and moon. The tide-raising power of the sun is about three-sevenths of that of the moon.

Tidal Friction. The friction caused by the motion of the tides. It has a tendency to check the speed of the the earth's rotation, but the effect is very small and only appreciable—if at all—after the lapse of ages.

Titan. The largest of Saturn's satellites, and sixth in order counting from the planet. It was discovered by C. Huygens on March 25th, 1655. Its mean distance from Saturn is about 777,000 miles, and it revolves round the planet in 15 days 22 hrs. 41 mins. It is visible in small telescopes, its stellar magnitude being, according to Pickering, 9·4. Its real diameter is somewhat doubtful, but is probably between 3,000 and 4,000 miles. It is therefore greater in volume than the planet Mercury.

Titania. One of the satellites of Uranus, the third in order of distance counting from the planet. Its mean distance from the planet's centre is about 291,000 miles, and its period of revolution 8 days 16 hrs. 56 mins. It can be well seen only in large telescopes, and its diameter is uncertain. Titania was discovered by Sir W. Herschel on Jan. 11th, 1787.

Total Eclipse. An eclipse of the sun in which the whole of the disc is covered by the moon; and an eclipse of the moon in which the moon is wholly immersed in the earth's shadow.

Toucan (the Toucan). One of the southern constellations. It lies south of Phœnix and Grus.

Trade Winds. These winds blow from the north-east in the northern hemisphere, and from the south-east in the southern hemisphere. They are due to air-currents flowing from the north and south towards the heated parts of the earth at the equator. They are deflected from their original course by the effect of the earth's rotation on its axis.

Transit. The passage of a celestial body across the meridian of the place of observation.

Transit Instrument. An instrument used for observing the passage of celestial bodies across the meridian. It consists of a telescope attached at right angles to a horizontal axis and fitted with vertical graduated circles. In the focus of the object-glass is a framework of cross wires.

Transit of a Satellite. The passage of a satellite across the disc of its primary planet.

Transit of a Shadow. The passage of the shadow of a satellite across the disc of a planet.

Transits of Mercury. The passage of the planet Mercury across the sun's disc. They occur more frequently than transits of Venus, but are not so useful for determining the sun's distance from the earth, owing to the proximity of Mercury to the sun, which renders the parallaxes of Mercury and the Sun more nearly equal than in the case of Venus and the Sun. Transits of Mercury at the same node occur at intervals of 7, 13, 33, or 46 years. The next transit will occur on Nov. 10th, 1894—the last of the present century.

Transits of Venus. The passage of the planet Venus across the sun's disc. They have been used to determine the sun's distance from the earth, but the results are not so satisfactory as might be expected. Transits of Venus occur at the following intervals in years : 8, 105½ ; 8, 121½ ; 8, 105½ ; 8, 121½. Transits took place in 1761, 1769, 1874, and 1882, and the next will occur in the years 2004 and 2012.

Transversal Disturbing Force. In the theory of perturbations, the component of the disturbing force, which acts at right angles to the radius vector, and in the same plane with the radius vector and the tangent to the orbit of the disturbed body.

Triangulum (the Triangle). One of the northern constellations.

Triangulum Australe. One of the southern constellations.

Tropical Revolution. The period of revolution of a planet, with reference to the nodes of its equator on the plane of its orbit.

Tropical Year. The time which elapses between two successive passages of the sun through the vernal equinox, or " First Point of Aries." The length of the tropical year is 365 days 5 hrs. 48 mins. 45·51 secs., or approximately 365¼ days.

Tropics. The two parallels or "small circles" on the earth's surface, which have a latitude north and south equal to the " obliquity of the ecliptic," or about 23° 27½'. The northern parallel is called the Tropic of Cancer, and the southern the Tropic of Capricorn. The region lying between these parallels is popularly spoken of as " the tropics."

True Sun. A term applied to the sun itself, to distinguish it from the imaginary or " mean sun."

Tureïs. A name sometimes applied to the star ι Argûs.

Twilight. The refracted sunlight visible after the sun has set, or before it rises. Twilight begins and ends when the sun is about 18° below the horizon. On the "longest day" the sun is about $23\frac{1}{2}°$ north of the celestial equator. Its zenith distance is therefore $66\frac{1}{2}°$; and if l be the latitude of the place, the sun's distance below the horizon at midnight will be $66\frac{1}{2}° - l$. Making this equal to 18°, we have $l = 66\frac{1}{2}° - 18 = 48\frac{1}{2}°$. Hence for all places on the earth's surface north of $48\frac{1}{2}°$ there is twilight all night on June 21st. North of latitude $66\frac{1}{2}°$ the sun does not set at all on the "longest day." This produces the phenomenon of "the midnight sun."

U.

Umbra. The dark shadow of the earth seen on the moon during a lunar eclipse. The umbra is bordered by a lighter shade called the *penumbra*. At points covered by the *umbra* a lunar spectator would see a total eclipse of the sun; but in the *penumbra* only a partial eclipse. The darker portion of a sun spot is called the *umbra*.

Umbriel. One of the satellites of Uranus, the second in order of distance counting from the planet. Its mean distance from the planet's centre is about 177,500 miles, and its period of revolution 4 days 3 hrs. 27 mins. It can be well seen only in large telescopes, and its diameter is uncertain. Umbriel was discovered by O. Struve on Oct. 8th, 1847.

Unukalhay. A name sometimes applied to the star a Serpentis. From the Arabic *nuk-al-hayyah*, "the serpent's neck."

Uranography. The department of astronomy which deals with the mapping of the stars.

Uranometry. The measurements of the heavens and of the positions of the fixed stars. The Latin term *Uranometria* has been applied to several star atlases. Thus, we have the *Uranometria Nova* of Argelander, Gould's *Uranometria Argentina*, etc.

Uranus. One of the superior planets. Discovered by Sir W. Herschel on March 13th, 1781. Its mean distance from the sun is about 1,780,000,000 miles, and its period of revolution about 84 years. Its diameter is about 33,000 miles. It may sometimes be seen with the naked eye, its stellar magnitude at opposition being about $5\frac{1}{2}$. Uranus is attended by four satellites—Ariel, Umbriel, Titania, and Oberon, which see.

Ursa Major (the Great Bear or " Plough "). One of the northern constellations.

Ursa Minor (the Little Bear). One of the northern constellations. Its principal star is Polaris, or the Pole star (a Ursæ Minoris).

V.

Variable Stars. Stars which are not constant in their light, but vary in brightness. Some of these curious and interesting objects vary to a great extent, but others only slightly. Over two hundred variable stars are now known. They have been arranged in the following classes : (1) temporary, or new stars, or *novæ* as they are also called; (2) variable stars, with long and tolerably regular periods; (3) irregular variables, or those which have no regular period, but fluctuate irregularly; (4) variables of short period; and (5) variables of the Algol, which at regular intervals undergo sudden diminutions of light, lasting for a few hours only. Of these classes, the following are examples: CLASS I. (temporary stars).—Tycho Brahé's star of 1572 in Cassiopeia, Kepler's Nova of 1604 in Ophiuchus, Schmidt's Nova Cygni in 1876, and Anderson's New Star in Auriga 1892, and some others (see TEMPORARY STARS). CLASS II.—Mira (o Ceti), χ Cygni, R. Leonis, etc. CLASS III.—a Herculis, a Orionis (Betelgeuse), μ Cephei, etc. CLASS IV.—β Lyræ, ζ Geminorum, η Aquilæ, δ Cephei, S. (10) Sagittæ, etc. CLASS V.—Algol, λ Tauri, δ Libræ. There are only ten known stars in this class.

Variation. An inequality in the moon's motion, due to the varying amount of the sun's disturbing force. This causes a maximum velocity of motion at "new moon" and "full moon," and a minimum velocity at the quadratures ("first" and "last quarter").

Vega. The bright star a Lyræ. It is sometimes spelt Wega. The name is derived from the Arabic *vaki.*

Vela. A name applied to a portion of the constellation Argo.

Velocity. The rate at which a body moves. This is usually expressed as so many feet per second; but in the case of very fast moving bodies, like the earth and planets, as so many miles per second.

Venus. One of the inferior planets, or those revolving round the sun inside the earth's orbit. Its mean distance from the sun is about 67,000,000 miles, and its orbit is more nearly circular than that of any of the other large planets. It revolves round the sun in a period of 224·7 days. Its diameter is about 7,918 miles, or nearly equal to that of the earth. As seen from the earth, Venus is the brightest of all the planets, and forms a brilliant object as a "morning" or "evening star." For further details see Appendix.

Vernal Equinox. The equinox at which the sun passes from the south to the north side of the ecliptic. This takes place about March 21st.

Vernier. "A short scale movable by the side of a longer scale, by which subdivisions of the longer scale may be measured." The longer scale is called the limb of the instrument. If the divisions on the vernier are shorter than those on the limb, the divisions on the vernier are numbered and read in the same direction as those on the limb, and the vernier is called a *direct vernier*. If the divisions on the vernier are longer than those on the limb, they are read in the opposite direction to those on the limb, and the vernier is called a *retrograde vernier*. (See *Engineers' Surveying Instruments*, by Professor Ira O. Baker, C.E.)

Vertex. The top of the disc of the sun, moon, or planets, or the point at which a great circle, passing through the zenith and the centre of the disc, intersects the limb.

Vertical Circles. Circles on the celestial sphere which pass through the zenith and nadir of the place of observation.

Vertical, Prime. The "great circle" on the celestial sphere, which passes through the zenith, nadir, and the east and west points of the horizon.

Vesta. One of the minor planets which revolve round the sun in orbits lying between those of Mars and Jupiter. It was discovered by Olbers on March 29th, 1807. It revolves round the sun in a period of 3·629 years, at a mean distance of 2·36 times the earth's mean distance from the sun. Vesta is the brightest of the group of minor planets, its magnitude at mean opposition being about 6½, and it has been occasionally seen with the naked eye. Its real diameter is probably about two hundred miles.

Via Lactea, or **Milky Way,** which see.

Vindemiatrix. A name sometimes applied to the star ϵ Virginis.

Virgo (the Virgin). One of the zodiacal constellations. Its brightest star is Spica (α Virginis).

Vis Viva. The mass of a moving body multiplied by the square of its velocity is called the *vis viva.*

Volans (the Flying Fish). One of the southern constellations.

Vulpecula (the Fox). One of the northern constellations.

W.

Waning Moon. The moon is said to be "waning" when its light is apparently decreasing between "full moon" and "new moon."

Wasat. A name sometimes applied to the star δ Geminorum.

Waxing Moon. The moon is said to be "waxing" between "new moon" and "full moon," when its light is apparently increasing.

Wedge Photometer. A form of photometer in which a wedge of tinted glass is used to extinguish the light of a star.

Wezen. A name sometimes applied to the star δ Canis Majoris.

Willow Leaves. A term applied by Nasmyth to the markings on the sun's surface, which he thought resembled in shape the leaves of the willow tree. They have also been termed "rice grains" and "granules."

Winter Solstice. The point on the ecliptic which is

8

at the maximum distance south of the equator. This point is reached by the sun about Dec. 22nd, which is popularly known as the "shortest day."

Y.

Year, Anomalistic. The time which elapses between two successive passages of the sun (in its apparent revolution among the stars) through the perigee of the earth's orbit. The length of the anomalistic year is 365 days 6 hrs. 13 mins. 48·09 secs.

Year, Civil. The year used for the ordinary affairs of life. It usually consists of 365 days, but as the real length of the year is about $365\frac{1}{4}$ days, a day is added every four years. This fourth year is called Leap Year, and contains 366 days.

Year, Leap. See preceding paragraph.

Year, Sidereal. The time which elapses between two successive returns of the sun to the same position among the fixed stars. The length of the sidereal year is 365 days 6 hrs. 9 mins. 9·314 secs. It is therefore about twenty minutes longer than the tropical year, and about four and a half minutes shorter than the anomalistic year.

Year, Synodic. A year of twelve lunar months, or about 355 days. It is a term not often used.

Year, Tropical. The time which elapses between two successive passages of the sun through the vernal equinox, or "First Point of Aries." The length of the tropical year is 365 days 5 hrs. 48 mins. 45·51 secs., or approximately $365\frac{1}{4}$ days.

Z.

Zaurac. A name sometimes applied to the star γ' Eridani.

Zavijava. A name sometimes applied to the star β Virginis.

Zenith. The point in the celestial sphere vertically overhead. Its direction is indicated by the plumb line.

Zenith Distance. The angular distance of a celestial body from the observer's zenith. It is the complement of the altitude, or the difference between 90° and the altitude.

Zenith Sector. An instrument for measuring the zenith distances of stars, invented by Hooke in 1669.

Zodiac. A belt of the sky extending along the ecliptic, in which the sun, moon, and most of the planets apparently perform their revolutions. The zodiacal zone is about 18° in width, 9° on each side of the ecliptic.

Zodiacal Light. "A cone-shaped or lenticular beam of light, which makes its appearance at certain times of the year above the eastern horizon in the mornings before dawn has commenced, and above the western horizon after sunset in the evening, remaining visible long after twilight has ceased" (*Astronomy for Amateurs*, p. 280). It is best seen after sunset in the spring months and before sunrise in the autumn. . In the tropics it is visible nearly every evening, and may occasionally be well seen in more northern latitudes. "The phenomenon is generally supposed to be due to a sort of nebulous envelope surrounding the sun, and densest in or near the ecliptic: hence the name of Zodiacal Light."

Zones. Spaces included between parallels of declination on the celestial sphere. The term is also applied to spaces on the earth's surface included between parallels of latitude, such as the "Torrid Zone," the "Temperate Zones," and "Frigid Zones."

Zosma. A name sometimes applied to the star δ Leonis. Derived from the Greek ζῶσμα, a tunic or girdle.

Zuben el Chameli. A name sometimes applied to the star β Libræ.

Zuben el Genubi. A name sometimes applied to the star α Libræ.

· **Zuben Hakrabi.** A name sometimes applied to the star γ Libræ.

APPENDIX.

ASTRONOMICAL DATA.

Coefficient of Refraction . . { 57·5″.

Horizontal Refraction . . { 33′.

Constant of Precession . . {
50·1882″ (Nyrén, 1869).
50·438239″ (Stockwell, 1873).
50·3514″ (L. Struve, 1888).

The limits of variation of this constant are, according to Stockwell, 48·212398″ and 52·664080″.

Constant of Nutation . . . {
9·23″ (Le Verrier, 1856).
9·134″ (E. J. Stone, 1869).
9·236″ (Nyrén, 1872).
9·22″ (Harkness, 1891).

Period of Nutation . . . { 18·66 years.

Equation of Equinoxes . { 15′ 37″.

Constant of Aberration . . {
20·4451″ (W. Struve, 1843).
20·492″ ± 0·006″ (Nyrén, 1883).
20·45451″ ± 0·01258″ (Harkness, 1891).
20·417″ ± 0·024″ (Lœwy and Puiseux, 1891).
20·494″ ± 0·017″ (Comstock, 1892).
20·510″ (Chandler, 1893).

Velocity of Light   186,337·0 ± 49·722 miles per sec. (Harkness).

Equation of Light   8 mins. 18 secs.

THE EARTH.

Equatorial Diameter $\left\{\begin{array}{l} \text{7926·59 miles (Clarke, 1880).} \\ \text{7926·248} \pm \text{0·156 miles (Harkness, 1891).} \end{array}\right.$

Polar Diameter . $\left\{\begin{array}{l} \text{7899·58 miles (Clarke, 1880).} \\ \text{7899·844} \pm \text{0·124 miles (Harkness, 1891).} \end{array}\right.$

Ellipticity or Compression . . $\left\{\begin{array}{l} \dfrac{1}{293\cdot47} \text{ (Clarke, 1878).} \\ \dfrac{1}{300\cdot205 \pm 2\cdot964} \text{ (Harkness, 1891).} \end{array}\right.$

Mean Density (water = 1) . . $\left\{\begin{array}{l} \text{5·66 (Francis Baily).} \\ \text{5·576} \pm \text{0·016 (Harkness, 1891).} \end{array}\right.$

Mass in Terms of Sun's Mass . . $\left\{ \dfrac{1}{327,214 \pm 624} \text{ (Harkness, 1891).} \right.$

	hrs.	mins.	secs.
Sidereal Day . .	23	56	4·1.
Mean Solar Day .	24	3	56·55.

	days	hrs.	mins.	secs.
Year, Tropical . .	365	5	48	46.
„ Sidereal . .	365	6	9	9·314 (Harkness, 1891).
„ Anomalistic .	365	6	13	48·09.

Eccentricity of the Earth's Orbit . $\left\{ \text{0·01677, or } \frac{1}{60} \text{ nearly.} \right.$

Obliquity of the Ecliptic . . $\left\{\begin{array}{l} 23° \ 27' \ 31\cdot83'' \text{ (Le Verrier, 1850).} \\ 23° \ 27' \ 22\cdot3'' \text{ (Airy, 1868).} \end{array}\right.$

Annual Motion of Line of Apsides . $\left\{ 11\cdot77''. \right.$

Acceleration of Gravity . . $\left\{ \text{32·086 feet per second (Harkness, 1891).} \right.$

Length of seconds Pendulum . . $\left\{ \text{39·012 inches.} \right.$

THE MOON.

Mean Parallax . . .	57' 2·542'' (Harkness, 1891).
Mean Distance from the Earth	$\left\{\begin{array}{l} \text{238,854·75} \pm \text{9·916 miles (Harkness).} \end{array}\right.$
Minimum Distance (perigee)	225.741·69 ± 5·44 miles.
Maximum Distance (apogee)	251,967·81 ± 5·44 miles.
Eccentricity of Orbit . .	0·05489972 (Harkness, 1891).
Inclination of Orbit . .	5° 8' 43·3546'' (Harkness).

Mean Angular Apparent Diameter . . .	{ 31' 5".
Diameter in Miles . .	2163.
Mass in Terms of Earth's Mass	{ $\frac{1}{81}$.
Density (water = 1) . .	3·40.
Force of Gravity (Earth's Gravity = 1) . . .	{ 0·165.
Albedo	0·174 (Zöllner).

	days	hrs.	mins.	secs.
Sidereal Month . . .	27	7	43	11·4.
Synodical Month . .	29	12	44	2·7.
Anomalistic Month . .	27	13	18	37·4.
Nodical Month . . .	27	5	5	35·8.

Period of Revolution of Moon's Nodes (sidereal) .	{ 6793·39 days.
Period of Revolution of Moon's Nodes (synodic) .	{ 346·644 days.
Period of Revolution of Moon's Apsides (sidereal)	{ 3232·575 days = 8·85 years.
Period of Revolution of Moon's Apsides (synodic)	{ 411·74 days.
Saros . . .	{ 223 Synodic Months = 6585·29 days = 18·09 years.
Metonic Cycle . .	{ 235 Synodic Months = 6939·69 days = 19 Tropical Years (very nearly).
Lunar Inequality of the Earth ...	6·52294" ± 0·01854" (Harkness).

THE SUN.

Solar Parallax .	{ 8·798" (Cornu). 8·79" ± 0·034" (Newcomb, 1890). 8·800" ± 0·03" (Auwers, 1891). 8·80906" ± 0·00567" (Harkness, 1891). 8·809 ± 0·0066 (Gill).
Sun's Mean Distance from the Earth . .	{ 92,796,950 ± 59,715 miles (Harkness, 1891).

Sun's Mean Angular Apparent Diameter .	$\begin{cases} 32' \; 3\cdot6'' \; (\textit{Nautical Almanack}). \\ 31' \; 59\cdot3'' \; (\text{Auwers}). \end{cases}$
Sun's Diameter in Miles . .	$\begin{cases} 866,000. \end{cases}$
Mass in Terms of Earth's Mass .	$\begin{cases} 327,214 \pm 624 \; (\text{Harkness}). \end{cases}$
Density (water = 1)	$1\cdot40.$
Force of Gravity at Sun's Equator (Earth's Gravity = 1). . .	$\begin{cases} \\ 27\cdot41. \\ \end{cases}$
Inclination of Sun's Equator to the Plane of the Ecliptic (1866·5)	$\begin{cases} \\ 6^\circ \; 58'. \\ \end{cases}$
Longitude of Ascending Node of Sun's Equator (1866·5) . .	$\begin{cases} \\ 74^\circ \; 36'. \\ \end{cases}$

MERCURY.

Mean Distance from the Sun (Earth's Distance = 1)	$\begin{cases} 0\cdot3870987. \end{cases}$
Mean Distance from the Sun in Miles .	35,921,579.
Eccentricity of Orbit	0·2056048.
Maximum Distance (aphelion) . .	43,308,000.
Minimum Distance (perihelion) . .	28,536,000.
Inclination of Orbit to Plane of Ecliptic	7° 0′ 8″.
Sidereal Period of Revolution . . .	87·969258 days.
Diameter in Miles	3000.
Polar Compression	$\frac{1}{20}$ (?).
Period of Rotation on Axis . . .	88 days (?).
Mass (Sun's Mass = 1) . . .	$\begin{cases} \dfrac{1}{8,374,672 \pm 1,765,762} \\ \text{(Harkness).} \end{cases}$
Mean Density (water = 1) . . .	4·00.
Force of Gravity at Equator (Earth's Gravity = 1)	$\begin{cases} 0\cdot272. \end{cases}$
Albedo	0·13 (Zöllner).

VENUS.

Mean Distance from the Sun (Earth's distance = 1)	0·7233322.
Mean Distance in Miles	67,123,022.
Eccentricity of Orbit	0·0068433.
Maximum Distance (aphelion)	67,582,364.
Minimum Distance (perihelion)	66,663,680.
Inclination of Orbit to Plane of Ecliptic	3° 23' 35".
Sidereal Period of Revolution	224·700787 days.
Diameter in Miles (0·999 of Earth)	7918 (Hartwig).
Polar Compression	Very small.
Period of Rotation on Axis	225 days (?).
Mass (Sun's Mass = 1)	$\dfrac{1}{408,968 \pm 1874}$ (Harkness).
Mean Density (water = 1)	4·46.
Force of Gravity at Equator (Earth's gravity = 1)	0·80.
Albedo	0·50 (Zöllner).

MARS.

Mean Distance from the Sun (Earth's distance = 1)	1·5236913.
Mean Distance from the Sun in Miles	141,393,905.
Eccentricity of Orbit	0·0932611.
Maximum Distance (aphelion)	154,580,456.
Minimum Distance (perihelion)	128,207,354.
Inclination of Orbit to Plane of Ecliptic	1° 51' 2".
Sidereal Period of Revolution	1·880832 year. 686·98 days.
Diameter in Miles	4200. 4700 (Niesten).
Polar Compression	About $\frac{1}{15}$, but uncertain.
Period of Rotation on Axis	24 hrs. 37 mins. 22·66 secs.
Mass (Sun's mass = 1)	$\dfrac{1}{3,093,500 \pm 3295}$ (Asaph Hall).
Mean Density (water = 1)	3·95.
Force of Gravity at Equator (Earth's gravity = 1)	0·376 (for diameter = 4200).
Albedo	0·2672 (Zöllner).

MINOR PLANETS.

No. 1. CERES.

Mean Distance from the Sun (Earth's distance = 1) . . . . } 2·767265.
Eccentricity of Orbit . . . 0·0763067.
Inclination of Orbit to Ecliptic . 10° 37′ 10″.
Sidereal Period of Revolution . . 4·603 years.
Diameter in Miles . . . . 196 (?).

No. 2. PALLAS.

Mean Distance from the Sun . . 2·767972.
Eccentricity of Orbit . . . 0·2408186.
Inclination of Orbit to Ecliptic 34° 43′ 55″ { Maximum Inclination of Group.
Sidereal Period of Revolution . . 4·605 years.
Diameter in Miles . . . . 171 (?).

No. 3. JUNO.

Mean Distance from the Sun . . 2·668256.
Eccentricity of Orbit . . . 0·2578570.
Inclination of Orbit to Ecliptic . 13° 1′ 23″.
Sidereal Period of Revolution . . 4·358 years.
Diameter in Miles . . . . 124 (?).

No. 4. VESTA.

Mean Distance from the Sun . . 2·361618.
Eccentricity of Orbit . . . 0·0884191.
Inclination of Orbit to Ecliptic . 7° 7′ 54″.
Sidereal Period of Revolution . . 3·629 years.
Diameter in Miles . . . . 214 (?).

No. 5. ASTRÆA.

Mean Distance from the Sun . . 2·578581.
Eccentricity of Orbit . . . 0·1863016.
Inclination of Orbit . . . . 5° 19′ 7″.
Sidereal Period of Revolution . . 4·141 years.
Diameter in Miles . . . . 57 (?).

No. 6. HEBE.

Mean Distance from the Sun	2·424993.
Eccentricity of Orbit	0·2034395.
Inclination of Orbit	14° 47′ 15″.
Sidereal Period of Revolution	3·776 years.
Diameter in Miles	92 (?).

No. 164. EVA.

Mean Distance from the Sun	2·631434.	
Eccentricity of Orbit	0·3471007	{ Maximum Eccentricity of Group.
Inclination of Orbit	24° 24′ 50″.	
Sidereal Period of Revolution	4·268 years.	

No. 279. THULE.

Mean Distance from the Sun	4·262060	{ Maximum Distance of Group.
Eccentricity of Orbit	0·0803782.	
Inclination of Orbit	2° 22′ 37″.	
Sidereal Period of Revolution	8·826 years.	

No. 149. MEDUSA.

Mean Distance from the Sun	2·174715	{ Minimum Distance of Group.
Eccentricity of Orbit	0·0707682.	
Inclination of Orbit	0° 55′ 12″.	
Sidereal Period of Revolution	3·207 years.	

No. 298. BAPTISTINE.

Mean Distance from the Sun	2·2198.
Eccentricity of Orbit	0·0000.
Inclination of Orbit	5° 0′.
Sidereal Period of Revolution	3·307 years.

JUPITER.

Mean Distance from the Sun (Earth's distance = 1)	$\left\{\right.$ 5·202800.
Mean Distance from the Sun in Miles	482,803,970.
Eecentricity of Orbit . . .	0·0482519.
Maximum Distance (aphelion). .	506,100,180 miles.
Minimum Distance (perihelion) .	459,507,760 miles.
Inclination of Orbit to Plane of Ecliptic	$\left\{\right.$ 1° 18′ 41″.
Sidereal Period of Revolution . .	11 years 314·838171 days.
Diameter in Miles . .	$\left\{\begin{array}{l}\text{Equatorial, } 89,790 \\ \text{Polar,} \quad\quad 84,300 \\ \quad\text{(Barnard).}\end{array}\right.$
Polar Compression	$\left\{\begin{array}{l}\frac{1}{17\cdot11}\ \text{(Kaiser).} \\ \frac{1}{15\frac{1}{2}}\ \text{(Schur).}\end{array}\right.$
Period of Rotation on Axis . .	9 hrs. 55 mins. 37 secs.
Mass (Sun's Mass = 1) . .	$\left\{\begin{array}{l}\dfrac{1}{1047\cdot55 \pm 0\cdot20} \\ \quad\text{(Harkness).}\end{array}\right.$
Mean Density (water = 1) . .	1·30.
Force of Gravity at Equator (Earth's gravity = 1)	$\left\{\right.$ 2·434.
Albedo	0·62 (Zöllner).

SATURN.

Mean Distance from the Sun (Earth's distance = 1)	$\left\{\right.$ 9·538861.
Mean Distance in Miles . . .	885,177,200.
Eecentricity of Orbit. . . .	0·0560713.
Maximum Distance (aphelion) . .	934,810,240.
Minimum Distance (perihelion) .	835,544,170.
Inclination of Orbit to Plane of Ecliptic	2° 29′ 40″.
Sidereal Period of Revolution . .	29 years 166·98636 days.
Diameter in Miles . . .	$\left\{\begin{array}{l}\text{Equatorial, } 75,900 \\ \text{Polar,} \quad\quad 67,600 \\ \quad\text{(Asaph Hall).}\end{array}\right.$
Polar Compression	$\left\{\right.\frac{1}{9\cdot18}\ \text{(Kaiser).}$

Period of Rotation on Axis . .	$\left\{\begin{array}{l}\text{10 hrs. 14 mins. 24 secs.}\\ \text{(Asaph Hall).}\end{array}\right.$
Mass (Sun's Mass = 1) . .	$\left\{\begin{array}{l}1\\ \overline{3501\cdot6\pm0\cdot78}\\ \text{(Harkness).}\end{array}\right.$
Mean Density (water = 1) . .	0·66.
Force of Gravity at Equator (Earth's Gravity = 1)	$\left\{\begin{array}{l}1\cdot02.\end{array}\right.$
Albedo	0·52 (Zöllner).

SATURN'S RINGS.

Exterior Diameter of Outer Ring .	173,500 miles.
Diameter of Ring in Middle of Cassini's Division	$\left\{\begin{array}{l}148,000\end{array}\right.$,,
Interior Diameter of Middle Ring .	112,400 ,,
Interior Diameter of Dusky Ring .	90,800 ,,
Width of Bright Rings . . .	30,500 ,,
Width between Dark Ring and Ball .	7600 ,,
Distance from Planet to Outside of Rings on the West	$\left\{\begin{array}{l}49,200\end{array}\right.$,,
Distance from Planet to Outside of Rings on the East	$\left\{\begin{array}{l}48,800\end{array}\right.$,,
Width of Cassini's Division about .	1700 ,,
Inclination of Rings to the Ecliptic .	28° 7′ 40″ (Asaph Hall).
Period of Rotation of Ring System .	$\left\{\begin{array}{l}\text{10 hrs. 32 mins. 15 secs.}\\ \text{(Sir W. Herschel).}\end{array}\right.$

[Mass of ring system $\frac{1}{720}$ of mass of Saturn according to Tisserand, but probably much less.]

URANUS.

Mean Distance from the Sun (Earth's distance = 1)	$\left\{\begin{array}{l}19\cdot18329.\end{array}\right.$
Mean Distance from the Sun in Miles .	1,780,150,800
Eccentricity of Orbit	0·0463402.
Maximum Distance (aphelion) . .	1,862,643,000.
Minimum Distance (perihelion) . .	1,697,658,600.
Inclination of Orbit to Plane of Ecliptic	0° 46′ 20″.
Sidereal Period of Revolution .	84 years 7·39036 days.

Diameter in Miles. . . . . 33,000.

Polar Compression . . . . $\frac{1}{71}$ (Schiaparelli).

Period of Rotation on Axis . . . Unknown.

Mass (Sun's mass $= 1$) . . . $\left\{ \dfrac{1}{22600 \pm 36} \right.$ (Harkness).

Mean Density (water $= 1$) . . . 1·11.

Force of Gravity at Equator (Earth's gravity $= 1$) . . . . . $\left\{ 0 \cdot 835. \right.$

Albedo . . . . 0·64 (Zöllner).

NEPTUNE.

Mean Distance from the Sun (Earth's distance $= 1$) . . . . $\left\{ 30 \cdot 05508. \right.$

Mean Distance from the Sun in Miles 2,789,019,700.

Eccentricity of Orbit . . . 0·0089646.

Maximum Distance (aphelion) . 2,814,022,000.

Minimum Distance (perihelion) . 2,764,017,000.

Inclination of Orbit to Plane of Ecliptic . . . . . $\left\{ 1° \ 47' \ 2''. \right.$

Sidereal Period of Revolution . . 164 years 280·11316 days.

Diameter in Miles . . . . 36,000.

Polar Compression . . . . Unknown.

Period of Rotation on Axis . . Unknown.

Mass (Sun's Mass $= 1$) . . . $\left\{ \dfrac{1}{18780 \pm 300} \right.$ (Harkness).

Mean Density (water $= 1$) . . 1·03.

Force of Gravity at Equator (Earth's Gravity $= 1$) . . . . $\left\{ 0 \cdot 844. \right.$

Albedo . . . . . . 0·46 (Zöllner).

SATELLITES OF MARS.

PHOBOS.

Mean Distance from Centre of Mars (Radius of Mars = 1) . .	2·771.
Mean Distance in Miles (Diameter of Mars = 4200 miles) . .	5819 miles.
Eccentricity of Orbit . . .	0·03208.
Inclination of Orbit . . .	26° 17·2′.
Sidereal Period of Revolution round Mars	7 hrs. 39 mins. 15·1 secs.
Diameter	About 7 miles.

DEIMOS.

Mean Distance from Centre of Mars (Radius of Mars = 1) .	6·921.
Mean Distance in Miles (Diameter of Mars = 4200) . . .	14,531 miles.
Eccentricity of Orbit . . .	0·00574.
Inclination of Orbit . . .	25° 47·2′.
Sidereal Period of Revolution round Mars	1 day 6 hrs. 17 mins. 54 secs.
Diameter	About 6 miles.

SATELLITES OF JUPITER.

	Barnard's Satellite.	I.	II.	III.	IV.
Mean Distance from Centre of Jupiter (Jupiter's radius = 1)	2·50	5·933	9·439	15·057	26·486
Mean Distance in miles from Centre of Jupiter (Jupiter's radius = 44,900 miles)	112,500	266,390	423,800	676,000	1,189,200
Eccentricity of Orbit	(?)	0·00	0·00	0·001316	0·007243
Inclination of Orbit	(?)	2° 8' 3"	1° 38' 57"	1° 59' 53"	1° 57' 0"
Sidereal Period of Revolution	h m s 11 57 23·06	d h m s 1 18 27 33·1	d h m s 3 13 13 42	d h m s 7 3 42 33·39	d h m s 16 15 32 11·2
Diameter in miles	100 ±	2400	2100	3430	2930
Mass (Jupiter's mass = 1)	(?)	0·000016877	0·000023227	0·000088437	0·000042175
Density (Water = 1)	(?)	1·12	2·14	1·87	1·47

SATELLITES OF SATURN.

	MIMAS.	ENCELADUS.	TETHYS.	DIONE.	RHEA.	TITAN.	HYPERION.	JAPETUS.
Mean Distance from Centre of Saturn (Saturn's equatorial radius = 1)	3·10	3·98	4·9	6·31	8·86	20·48	25·07	59·58
Mean Distance in Miles (Saturn's equatorial radius = 37,950)	117,645	151,410	187,600	230,460	336,240	777,200	951,400	2,261,000
Eccentricity of Orbit	0·016	0·0047	—	0·00396	0·00364	0·029869	·11885	0·09257
Inclination of Orbit	27° 36'	28° 7' 0"	28° 40' 12"	27° 58' 36"	27° 54' 27"	27° 38' 49"	27° 4' 8"	18° 31' 5"
Sidereal Period of Revolution	d h m s 0 22 37 5·1	d h m s 1 8 53 7	d h m s 1 21 18 26	d h m s 2 17 41 0·4	d h m s 4 12 25 11·6	d h m s 15 22 41 23	d h m s 21 6 39 27	d h m s 79 7 54 17
Diameter in Miles	—	—	—	—	1200 ?	3200 ?	—	1800 ?
Mass (Saturn's mass = 1) all uncertain	0·000,000,09	0·000,000,25	0·000,001,30	0·000,001,59	—	387 × Mass of Hyperion (Asaph Hall).	—	0·000,010

9

SATELLITES OF URANUS.

	Ariel.	Umbriel.	Titania.	Oberon.
Mean Distance from Centre of Uranus (Equatorial radius of Uranus = 1)	7·72	10·76	17·65	23·60
Mean Distance in Miles (Radius of Uranus = 16,500)	127,380	177,540	291,225	389,400
Eccentricity of Orbit	0·020	0·010	0·00106	0·00383
Inclination of Orbit	97° 58′	98° 21′	97° 47′	97° 54′
Sidereal Period of Revolution	$2^d\ 12^h\ 29^m\ 21^s$	$4^d\ 3^h\ 27^m\ 37^s$	$8^d\ 16^h\ 56^m\ 29{\cdot}5^s$	$13^d\ 11^h\ 7^m\ 6{\cdot}4^s$

SATELLITE OF NEPTUNE.

Mean Distance from Centre of Neptune (Radius of Neptune = 1) . . . . 14·54.
Mean Distance in Miles (Radius of Neptune = 18,000) . . . . 261,700.
Eccentricity of Orbit . . . . 0·0088.
Inclination of Orbit . . . . 145°.
Sidereal Period of Revolution . . . . $5^d\ 21^h\ 2^m\ 44^s$.

LIST OF REMARKABLE RED STARS.

(Only the very reddest stars and those brighter than the ninth magnitude are included.)

STAR	R.A. 1890. hrs. mins. secs.	DECL. 1890. degrees mins.	MAG.	REMARKS.
Birmingham 1	0 11 5	+ 41 59	8·2	"Almost vermilion" (Franks); "Intense red colour, most wonderful" (Espin).
D.M. +31°, 56	0 21 43	+ 34 59·6	8·1	"Presque rouge absolu" (Dunér).
R Sculptoris .	1 21 54	− 33 7·0	Var.	"Intense scarlet" (Gould).
"Mira (ο) Ceti"	2 13 47	− 3 28·6	Var.	Very red at minimum.
Birm. 65 .	3 32 21	+ 62 17·5	7·3	"Fiery red" (Dreyer).
R Doradûs .	4 35 29	− 62 17·6	Var.	"Very red" (Thome).
Birm. 85 .	4 44 37	− 28 20·3	8	"Extraordinary ruby colour" (Sir J. Herschel); "Very red" (Espin).
D.M. +38°, 955 .	4 45 6	+ 38 18·9	8·8	"Very red" (Espin).
R Leporis .	4 54 36	− 14 58·3	Var.	"Most intense crimson" (Hind).
Birm. 96 .	4 59 43	+ 1 1·5	6·6	"Fiery red" (Doberck); "Fine ruby" (Webb).
D.M. +7°, 929 .	5 27 17	+ 7 3·8	8·2	"Very red" (Espin).
Birm. 120 .	5 38 29	+ 21 22·3	8·5	"Presque rouge absolu" (Dunér).
Birm. 121 .	5 39 6	+ 20 38·9	7·7	"Splendid crimson" (Birmingham); "Full red" (Copeland); "Orange red" (Espin).
Birm. 135 .	6 4 3	+ 26 2·1	7·4	"Fine ruby colour" (Webb); "Deep red orange" (Gemmill).
Birm. 148 .	6 28 59	+ 38 32·0	6·3	"Splendid red" (Dreyer); "Colour wonderful" (Espin).
Birm. 165 .	7 1 47	− 7 23·3	8·3	"Remarkably fine red" (Espin).
S Canis Minoris .	7 26 45	+ 8 33·0	Var.	"Fiery red" (Hind).

STAR.	R. A. 1880.			DECL. 1880.		MAG.	REMARKS.
	hrs.	mins.	secs.	degrees	mins.		
R Leonis	9	41	39	+ 11	56·3	Var.	"Blood red" (Criswick); "Presque rouge absolu" (Dunér).
Birm. 232	9	57	34	− 59	45·8	7¾	"Scarlet" (Sir J. Herschel); "Very red" (Thome).
V Hydræ	10	46	17	− 20	40·0	Var.	"Copper red, most magnificent" (Dreyer).
R Crateris	10	55	9	− 17	44·1	Var.	"Scarlet" (Winnecke); "Very intense ruby" (Webb).
Birm. 225	11	5	30	− 81	11·9	8¾	"Ruby" (Sir J. Herschel).
D.M. + 56°, 1615	12	35	21	+ 56	26·7	8·2	"Very red" (Espin).
Birm. 291	12	40	59	− 59	5·6	9·0	"Most intense blood red" (Sir J. Herschel). In field with β Crucis.
Birm. 313	13	42	49	− 27	49	7·0	"Deep red or crimson" (Burnham); "Very red" (Espin).
D.M. + 33°, 2482	14	34	39	+ 33	0·5	8·2	"Fine red" (Espin).
Birm. 347	15	14	58	− 75	32·3	8¾	"Very high red" (Sir J. Herschel).
V Ophinchi	16	20	36	− 12	10·6	Var.	"Genuine ruby" (Birmingham); "Presque rouge absolu" (Dunér).
Birm. 385	16	33	34	− 32	9·8	9	"Deep red, like a drop of blood" (Sir J. Herschel).
Birm. 396	16	53	33	− 54	54·4	8½	"Intense ruby red" (Sir J. Herschel).
Birm. 410	17	23	14	− 19	23·0	7·8	"Fine ruby" (Birmingham); "Intense red" (Copeland); "Very red" (Espin).
Birm. 418	17	38	29	− 18	36·5	8·5	"Remarkable red" (Sir J. Herschel); "Very intense red" (Burton).
Birm. 448	18	28	32	+ 36	54·6	8·5	"Intense" (Secchi); "Fiery red, superb" (Franks); "Crimson, magnificent" (Espin); "Presque rouge absolu" (Dunér).
D.M. + 8°, 3780	18	33	6	+ 8	44·0	Var.	"Fine red" (Espin).

Star.	R. A. 1890.			Decl. 1890.		Mag.	Remarks.
	hrs.	mins.	secs.	degrees	mins.		
Birm. 464	18	43	57	− 8	1·8	7·1	"Most remarkable red" (Sir J. Herschel); "Very fine red" (Birmingham); "Very red" (Espin).
Birm. 475	18	53	31	+11	2·9	9·0	"Fine red" (Birmingham); "Fiery red" (Copeland); "Very red" (Espin); "Presque rouge absolu" (Dunér).
Birm. 483	18	58	32	− 5	50·8	7·0 Var.?	"Truly striking and wonderful" (Webb); "Very red" (Espin); "Presque rouge absolu" (Dunér).
Birm. 521	19	53	38	+43	57·7	8·2	"Splendid, like a drop of blood" (Franks).
Birm. 545	20	10	40	−21	38·3	7·7	"Pure red— perhaps the finest of my ruby stars" (Sir J. Herschel); "Very red" (Dreyer).
U Cygni	20	16	12	+47	32·8	Var.	"One of the loveliest hues in the sky" (Webb); "Very red" (Espin).
D.M. +39°, 4152	20	17	40	+40	5·7	8·0	"Very red" (Espin). In field with γ Cygni.
V Cygni	20	37	46	+47	44·9	Var.	"Very red" (Birmingham); "Presque rouge absolu" (Dunér).
D.M. +45°, 3349	20	54	11	+46	2·7	8·1	"Very red" (Espin).
S Cephei	21	36	35	+78	7·7	Var.	"Very deep red" (Copeland); "Presque rouge absolu" (Dunér).
Birm. 592	21	38	43	+37	30·8	Var.	"Splendid red" (Birmingham); "Very fine colour" (Webb); "Orange vermilion" (Franks); "Presque rouge absolu" (Dunér).
μ Cephei	21	40	8	+58	16·5	Var.	Herschel's "garnet star." Probably the reddest star visible to the naked eye.
Birm. 658	23	55	39	+59	44·6	7·8	"A very fine ruby, intense and beautiful, pure red" (Webb); "Presque rouge absolu" (Dunér).

LIST OF REMARKABLE VARIABLE STARS.

(*Only those visible, or sometimes visible, to the naked eye are given.*)

STAR.	R.A. 1890. hrs, mins.	DECL. 1890. degrees, mins.	VARIATION Maximum	VARIATION Minimum	REMARKS.
T Ceti .	0 16·2	− 20 40	5·1 – 5·3	6·4 – 7·0	Irregular.
B Cassiopeiæ .	0 18·7	+ 63 32	> 1	(?)	Nova, 1572.
O ("Mira") Ceti .	2 13·8	− 3 28·6	1·7 – 5·0	8 – 9·5	Mean period 331·33 days.
ρ Persei .	2 58·1	+ 38 25	3·4	4·2	Irregular.
β Persei (Algol) .	3 1·0	+ 40 32	2·3	3·5	Type of Algol variables. Period 2 days 20 hrs. 48 mins. 51 secs., from minimum to minimum.
λ Tauri .	3 54·6	+ 12 11	3·4	4·2	Algol type. Period 3 days 22 hrs. 52 mins. 12 secs.
Nova Aurigae .	5 25·0	+ 30 21·8	4·5	> 15	Nova, 1892.
α Orionis .	5 49·2	+ 7 23	1	1·4	Irregular.
U Orionis . .	5 49·3	+ 20 9·3	6 – 7·5	> 12	"Nova," 1885. Period 373½ days.
η Geminorum .	6 8·2	+ 22 32	3·2	3·7 – 4·2	Period 229 days.
T Monocerotis .	6 19·3	+ 7 8·7	5·8 – 6·4	7·4 – 8·2	„ 27 days.
ζ Geminorum .	6 57·6	+ 20 44	3·7	4·5	„ 10 days 3 hrs. 41·5 mins.
L₂ Puppis. .	7 10·2	− 44 28	3·5	6·3	„ 137 days.
R Canis Majoris	7 14·5	− 16 11	5·9	6·7	Algol type. Period 1 day 3 hrs. 15 mins. 46 secs.
R Carinæ .	9 29·5	− 62 18	4·3 – 5·7	9·3 – 10·0	Period 312 days.

Star.	R.A. 1890. hrs. mins.	Decl. 1890. degrees mms.	Variation. Maximum	Variation. Minimum	Remarks.
R Leonis	9 41·6	+ 11 56·3	5·2 — 6·7	9·1 — 10·0	Period 313 days.
l Carinæ	9 42·2	— 62 0	3·7	5·2	„ 31 days.
R Ursæ Majoris	10 36·9	+ 69 21	6·0 — 8·2	13	Mean period 302 days.
η Argûs	10 40·8	— 59 6	> 1	7·6	Irregular.
R Hydræ	13 23·7	— 22 43	3·5 — 5·5	9·7	Period 437 days, decreasing.
S Virginis	13 27·3	— 6 38	5·7 — 7·8	12·5	Period 376 days.
R Centauri	14 8·7	— 59 24	5·6 — 6·3	8·7 — 9·8	„ about 530 days.
R Boötis	14 32·4	+ 27 13	5·9 — 7·8	11·3 — 12·2	„ 223·9 days.
W (34) Boötis	14 38·6	+ 27 0	5·2	6·1	Long and irregular.
δ Libræ	14 55·1	— 8 5	5·0	6·2	{ Algol type. Period 2 days 7 hrs. 51 mins. 23 secs.
R Coronæ	15 44·1	+ 28 30	5·8	13·0	Irregular.
R Serpentis	15 45·6	+ 15 28	5·6 — 7·6	13	Period 357·6 days.
T Coronæ	15 54·9	+ 26 14	2·0	9·5	Nova, 1866. "The Blaze Star."
g (30) Herculis	16 25·0	+ 42 8	4·7 — 5·5	5·4 — 6·0	Irregular.
S Herculis	16 46·9	+ 15 8	5·9 — 7·5	11·5 — 13	Period 309·2 days.
Ophiuchi	16 53·3	— 12 43·5	5·5	12·5	Nova, 1848.
α Herculis	17 9·6	+ 14 31	3·1	3·9	Irregular.
U Ophiuchi	17 10·9	+ 1 20	6·0	6·7	{ Algol type. Period 20 hrs. 7 mins. 41 secs.

Star.	R.A. 1890		Decl. 1890		Variation.		Remarks.
	hrs.	mins.	degrees	mins.	Maximum.	Minimum.	
u (68) Herculis	17	13.3	+33	13	4.6	5.4	
—Serpentarii	17	24.1	−21	23.3	>1	(?)	Kepler's nova, 1604.
X Sagitarii	17	40.6	−27	47	4	6	Period 7.011 days.
W Sagittarii	17	58.0	−29	35	5	6.5	„ 7.593 days.
Y Sagittarii	18	14.9	−18	54	5.8	6.6	„ 5.769 days.
R Scuti	18	41.6	−5	50	4.7 − 5.7	6.0 − 9.0	Mean period 71.1 days.
κ Pavonis	18	45.6	−67	22	4.0	5.5	Period 9.1014 days.
β Lyræ	18	46.1	+33	14	3.4	4.5	„ 12days 21hrs. 46mins. 58.3 secs.
Nova Vulpeculæ	19	43.1	+27	2.6	3	(?)	Nova, 1670.
χ Cygni	19	46.3	+32	38	4.0 − 6.5	13.5	Period 406 days.
η Aquilæ	19	46.9	+0	43	3.5	4.7	„ 7 days 4 hrs. 14 mins.
S (10) Sagittæ	19	51.0	+16	21	5.6	6.4	„ 8 days 9 hrs. 11 mins.
P (34) Cygni	20	13.7	+37	41	3 − 5	>6	Nova, 1600.
T Vulpeculæ	20	46.6	+27	49	5.5	6.5	Period 4 days 10 hrs. 29 mins.
T Cephei	21	8.0	+68	2.6	5.6 − 6.8	9.5 − 9.9	„ 383.2 days.
Cygni	21	37.4	+42	20.4	3	13.5	Nova, 1876.
δ Cephei	22	25.1	+57	51	3.7	4.9	Period 5 days 8 hrs. 47 mins. 40 secs.
β Pegasi	22	58.4	+27	29	2.2	2.7	Irregular.
R Aquarii	23	38.1	−15	53	5.8 − 8.5	11 (?)	Period 387 days.
R Cassiopeiæ	23	52.8	+50	46.5	4.8 − 7.0	9.8 − 12	„ 429 days.

LIST OF BINARY STARS FOR WHICH ORBITS HAVE BEEN COMPUTED.

STAR	R.A. 1880. hrs.	mins.	secs.	DECL. 1880. degrees	mins.	PERIOD Years.	REMARKS.
Struve 3062	0	0	30	+57	49·4	102·9	Orbit by Doberek, 1879.
O. Struve 4	0	11	0	+35	52·2	135·2	„ Glasenapp, 1889.
η Cassiopeiæ	0	42	27	+57	13·9	167·4	„ Coit, 1882.
66 Piscium	0	48	50	+18	35·5	136·2	„ Glasenapp, 1889.
36 Andromedæ	0	49	5	+23	1·9	316·0	„ Doberek, 1875.
ρ Eridani	1	35	36	−56	45	302·37	„ computed, 1887.
Struve 186	1	50	18	+1	19	150·80	by Glasenapp, 1891.
Struve 228	2	7	0	+46	59	88·73	computed, 1889.
40 (O²) Eridani (BC)	4	10	13	−7	49·5	139·0	requires revision
Burnham 883	4	45	6	+10	53	16·35	by Glasenapp, 1892.
14 (ι) Orionis	5	1	54	+8	21	190·48	computed, 1887.
O. Struve 119	6	29	34	+29	22	85·9	by Glasenapp, 1889.
12 Lyncis (AB)	6	36	31	+59	33	485·8	computed, 1887.
Sirius	6	40	18	−16	34·0	49·4	by Auwers, 1892.
Castor	7	27	35	+32	7·7	1001·21	„ Doberek.
9 Argûs	7	46	12	+13	36	40·54	„ Glasenapp, 1892.
ξ Cancri (AB)	8	5	51	−17	58·7	59·11	„ Seeliger, 1888.
Struve 3121	9	11	21	+29	2·4	34·64	„ Celoria, 1887.
ι Leonis	9	22	34	+9	32·1	115·30	„ Doberek, 1876.
φ Ursæ Majoris	9	44	37	+54	35·6	115·4	„ Casey, 1882.
δ Sextantis	9	47	6	−7	35	93·92	„ Glasenapp, 1892.
O. Struve 215	10	10	14	+18	18	107·94	computed, 1890.
γ Leonis	10	13	54	+20	23·8	407·0	„ by Doberek, 1879.

STAR.	R. A. 1890.			DECL. 1890.		PERIOD YEARS.	REMARKS.
	hrs.	mins.	secs.	degrees	mins.		
ξ Ursæ Majoris	11	12	19	+32	8·9	60·8	Orbit by Pritchard, 1878.
ι Leonis	11	18	12	+12	8	116·27	„ computed, 1891.
O. Struve 234	11	24	53	+41	54·9	63·45	„ „ 1886.
O. Struve 235	11	26	6	+61	40·5	94·4	„ by Doberek, 1879.
γ Centauri	12	35	27	−48	21·3	61·88	„ computed, 1892.
γ Virginis	12	36	5	0	50·8	180·54	„ well determined, Doberek, 1881.
35 Comæ	12	47	54	+21	51	228·42	„ computed, 1891.
42 Comæ	13	4	38	+18	6·7	25·71	„ by O. Struve.
O. Struve 269	13	27	53	+35	28	47·70	„ computed, 1892.
Struve 1757	13	28	41	0	14·9	276·92	„ „ 1892.
25 Can. Venat.	13	32	34	+36	51·3	119·92	„ by Doberek, 1880.
Burnham 612	13	34	12	+11	18	30·00	„ „ Glasenapp, 1892.
Struve 1785	13	44	6	+27	32	125·52	„ computed, 1893.
Struve 1819	14	9	48	+3	38·2	340·1	„ by Casey, 1882.
α Centauri	14	32	10	−60	22·8	88·5	„ provisional.
ξ Boötis	14	16	19	+19	33·5	127·35	„ by Doberek, 1877.
44 (i) Boötis	15	0	10	+48	5	261·12	„ „ „ 1875.
η Coronæ Bor.	15	18	40	+30	41·2	41·56	„ „ „ 1880.
μ² Boötis	15	20	20	+37	44·0	280·29	„ „ „ 1878.
O. Struve 298	15	32	3	+40	11·6	56·65	„ by Celoria, 1888.
γ Coronæ Bor.	15	38	7	+26	38·8	95·5	„ „ Doberek, 1877.
ξ Scorpii	15	58	19	−11	4·1	105·19	„ „ Schorr, 1889.
σ Coronæ Bor.	16	10	33	+34	9·3	845·86	„ „ Doberek, 1876.
λ Ophiuchi	16	25	22	+2	13·6	373.5	„ „ Glasenapp, 1888.
ζ Herculis	16	37	8	+31	48·2	31·411	„ „ Doberek, 1880.

STAR	R.A. 1890 (hrs, mins, secs)			DECL. 1890 (degrees, mins)		PERIOD Years	REMARKS
Struve 2091?	16	40	32	+43	41·5	205·46	Orbit computed, 1882.
Struve 2107	16	47	29	+28	51·5	186·2	„ by Berberich, 1884.
μ Draconis	17	3	3	+51	37·0	6180	„ „ 1884.
Burnham 416	17	10	47	−34	61·2	34·48	„ „ computed, 1893.
Struve 2173	17	21	41	0	58·2	45·43	„ „ by Dunér, 1876.
μ′ Herculis	17	42	7	+27	47·2	45·39	„ „ Leuehner, 1889.
η Ophiuchi	17	57	6	−8	107	217·87	„ „ Doberck, 1875.
70 Ophiuchi	17	59	61	+2	31·6	87·84	„ „ Gore, 1888.
99 Herculis	18	2	51	+30	33	53·55	„ „ 1890.
ζ Sagittarii	18	55	37	−30	2·2	18·69	„ „ 1886.
γ Coronæ Aust.	18	58	59	−37	13·2	154·41	„ „ 1892.
Struve 2525	19	22	8	+27	9	138·54	„ „ 1892.
δ Cygni	19	41	32	−11	51·8	376·66	„ „ 1890.
O. Struve 387	19	41	38	+35	2	110·1	„ „ Glasenapp, 1889.
O. Struve 400	20	6	35	+43	37	170·37	„ „ Gore, 1887.
β Delphini	20	32	33	+11	12·8	30·9	„ „ 1885.
λ Cygni	20	43	6	+36	6	93·4	„ „ Glasenapp, 1889.
4 Aquarii	20	45	26	+6	2·3	129·84	„ „ Doberek, 1880.
61 Cygni	21	1	58	−38	12·5	462·0	Binary character doubtful.
δ Equulei	21	9	6	+9	34	11·478	Shortest period known.
κ Pegasi	21	39	42	+25	8	11·54	Orbit by Glasenapp.
ξ Aquarii	22	23	10	0	32	1578·33	Longest period known, Doberek, 1875.
37 Pegasi	22	21	24	+3	52	117·54	Orbit by Gore, 1892.
π Cephei	23	4	24	+74	47·6	198·4	„ „ Glasenapp, 1889.
85 Pegasi	23	56	26	+26	300	17·487	„ „ 1892.

THE HEAVENS AND THEIR ORIGIN.

THE VISIBLE UNIVERSE : Chapters on the Origin
and Construction of the Heavens. By J. E. GORE, F.R.A.S., etc.
Illustrated by Six Stellar Photographs and Twelve Photographic
Plates. Demy 8vo, 16s. cloth, gilt top.

" A valuable and lucid summary of recent astronomical theory, rendered more
valuable and attractive by a series of stellar photographs and other illustrations."—
h. Times.

" In presenting a clear and concise account of the present state of our knowledge,
Mr. Gore has made a valuable addition to the literature of the subject."—*Nature.*

"Mr. Gore's ' Visible Universe ' is one of the finest works on astronomical science
that has recently appeared in our language. In spirit and in method it is scientific
from cover to cover, but the style is so clear and attractive that it will be as
acceptable and as readable to those who make no scientific pretensions as to those
who devote themselves specially to matters astronomical."—*Leeds Mercury.*

" We are glad to bear witness to the fulness, the accuracy, and the entire honesty
of the latest and the best compilation of the kind which has appeared of late years.
. . . The illustrations also are admirable."—*Daily Chronicle.*

" As interesting as a novel, and instructive withal ; the text being made still
more luminous by stellar photographs and other illustrations. . . . A most valuable
book."—*Manchester Examiner.*

THE CONSTELLATIONS.

STAR GROUPS : A Student's Guide to the Constellations.
By J. E. GORE, F.R.A.S., etc. With Thirty Maps. Small 4to,
5s. cloth, silvered.

☞ *The Maps are intended as an aid to the beginner in acquiring a knowledge
of the principal constellations. All stars to the sixth magnitude are shown,
this being about the limit of ordinary eyesight. A letterpress explanation
is added to each map, giving some account of the most interesting objects
in each constellation. It is hoped that these little maps will be found useful
as an introduction to larger atlases and more extensive works on the subject.*

" A knowledge of the principal constellations visible in our latitudes may be
easily acquired from the thirty maps and accompanying text contained in thi
work."—*Nature.*

"The volume contains thirty maps showing stars of the sixth magnitude—the
usual naked-eye limit—and each is accompanied by a brief commentary, adapted
to facilitate recognition and bring to notice objects of special interest. For the
purpose of a preliminary survey of the 'midnight pomp' of the heavens, nothing
could be better than a set of delineations averaging scarcely twenty square inches
in area, and including nothing that cannot at once be identified."—*Saturday
Review.*

"A very compact and handy guide to the constellations."—*Athenæum.*

LONDON:
CROSBY LOCKWOOD & SON, 7, STATIONERS' HALL COURT, E.C.

7, STATIONERS' HALL COURT, LONDON, E.C.

October, 1891.

A

CATALOGUE OF BOOKS

INCLUDING NEW AND STANDARD WORKS IN

ENGINEERING: CIVIL, MECHANICAL, AND MARINE, MINING AND METALLURGY, ELECTRICITY AND ELECTRICAL ENGINEERING, ARCHITECTURE AND BUILDING, INDUSTRIAL AND DECORATIVE ARTS, SCIENCE, TRADE AGRICULTURE, GARDENING, LAND AND ESTATE MANAGEMENT, LAW, &c.

PUBLISHED BY

CROSBY LOCKWOOD & SON.

MECHANICAL ENGINEERING, etc.

New Pocket-Book for Mechanical Engineers.

THE MECHANICAL ENGINEER'S POCKET-BOOK OF TABLES, FORMULÆ, RULES AND DATA. A Handy Book of Reference for Daily Use in Engineering Practice. By D. KINNEAR CLARK, M.Inst.C.E., Author of "Railway Machinery," "Tramways," &c. &c. Small 8vo, nearly 700 pages. With Illustrations. Rounded edges, cloth limp, 7s. 6d.; or leather, gilt edges, 9s. [*Just published.*

New Manual for Practical Engineers.

THE PRACTICAL ENGINEER'S HAND-BOOK. Comprising a Treatise on Modern Engines and Boilers: Marine, Locomotive and Stationary. And containing a large collection of Rules and Practical Data relating to recent Practice in Designing and Constructing all kinds of Engines, Boilers, and other Engineering work. The whole constituting a comprehensive Key to the Board of Trade and other Examinations for Certificates of Competency in Modern Mechanical Engineering. By WALTER S. HUTTON, Civil and Mechanical Engineer, Author of "The Works' Manager's Handbook for Engineers," &c. With upwards of 370 Illustrations. Third Edition, Revised, with Additions. Medium 8vo, nearly 500 pp., price 18s. Strongly bound.

☞ *This work is designed as a companion to the Author's* "WORKS' MANAGER'S HAND-BOOK." *It possesses many new and original features, and contains, like its predecessor, a quantity of matter not originally intended for publication, but collected by the author for his own use in the construction of a great variety of modern engineering work.*

. OPINIONS OF THE PRESS.

" A thoroughly good practical handbook, which no engineer can go through without learning something that will be of service to him."—*Marine Engineer.*

" An excellent hook of reference for engineers, and a valuable text-book for students of ngineering."—*Scotsman.*

" This valuab'e manual embodies the results and experience of the leading authorities on mechanical engineering."—*Building News.*

" The author has collected together a surprising quantity of rules and practical data, and has shown much judgment in the selections he has made. . . . There is no doubt that this book is one of the most useful of its kind published, and will be a very popular compendium."—*Engineer.*

" A mass of information, set down in simple language, and in such a form that it can be easily referred to at any time. The matter is uniformly good and well chosen, and is greatly elucidated by the illustrations. The book will find its way on to most engineers shelves, where it will rank as one of the most useful books of reference.'—*Practical Engineer.*

" Should be found on the office shelf of all practical engineers."—*English Mechanic.*

B

Handbook for Works' Managers.

THE WORKS' MANAGER'S HANDBOOK OF MODERN RULES, TABLES, AND DATA. For Engineers, Millwrights, and Boiler Makers; Tool Makers, Machinists, and Metal Workers; Iron and Brass Founders, &c. By W. S. HUTTON, C.E., Author of "The Practical Engineer's Handbook." Fourth Edition, carefully Revised, and partly Re-written. In One handsome Volume, medium 8vo, 15s. strongly bound. [*Just published.*

☞ *The Author having compiled Rules and Data for his own use in a great variety of modern engineering work, and having found his notes extremely useful, decided to publish them—revised to date—believing that a practical work, suited to the* DAILY REQUIREMENTS OF MODERN ENGINEERS, *would be favourably received.*

In the Third Edition, the following among other additions have been made, viz.: Rules for the Proportions of Riveted Joints in Soft Steel Plates, the Results of Experiments by PROFESSOR KENNEDY *for the Institution of Mechanical Engineers—Rules for the Proportions of Turbines—Rules for the Strength of Hollow Shafts of Whitworth's Compressed Steel, &c.*

*** OPINIONS OF THE PRESS.

"The author treats every subject from the point of view of one who has collected workshop notes for application in workshop practice, rather than from the theoretical or literary aspect. The volume contains a great deal of that kind of information which is gained only by practical experience, and is seldom written in books."—*Engineer.*

"The volume is an exceedingly useful one, brimful with engineers' notes, memoranda, and rules, and well worthy of being on every mechanical engineer's bookshelf."—*Mechanical World.*

"The information is precisely that likely to be required in practice. . . . The work forms a desirable addition to the library not only of the works manager, but of anyone connected with general engineering."—*Mining Journal.*

"A formidable mass of facts and figures, readily accessible through an elaborate index . . . . Such a volume will be found absolutely necessary as a book of reference in all sorts of 'works' connected with the metal trades."—*Ryland's Iron Trades Circular.*

"Brimful of useful information, stated in a concise form, Mr. Hutton's books have met a pressing want among engineers. The book must prove extremely useful to every practical man possessing a copy."—*Practical Engineer.*

Practical Treatise on Modern Steam-Boilers.

STEAM-BOILER CONSTRUCTION. A Practical Handbook for Engineers, Boiler-Makers, and Steam Users. Containing a large Collection of Rules and Data relating to the Design, Construction, and Working of Modern Stationary, Locomotive, and Marine Steam-Boilers. By WALTER S. HUTTON, C.E., Author of "The Works' Manager's Handbook," &c. With upwards of 300 Illustrations. Medium 8vo, 18s. cloth. [*Just published.*

"Every detail, both in boiler design and management, is clearly laid before the reader. The volume shows that boiler construction has been reduced to the condition of one of the most exact sciences ; and such a book is of the utmost value to the *fin de siècle* Engineer and Works' Manager."—*Marine Engineer.*

"There has long been room for a modern handbook on steam boilers; there is not that room now, because Mr. Hutton has filled it. It is a thoroughly practical book for those who are occupied in the construction, design, selection, or use of boilers."—*Engineer.*

"The Modernised Templeton."

THE PRACTICAL MECHANIC'S WORKSHOP COMPANION. Comprising a great variety of the most useful Rules and Formulæ in Mechanical Science, with numerous Tables of Practical Data and Calculated Results for Facilitating Mechanical Operations. By WILLIAM TEMPLETON, Author of "The Engineer's Practical Assistant," &c. &c. Sixteenth Edition, Revised, Modernised, and considerably Enlarged by WALTER S. HUTTON, C.E., Author of "The Works' Manager's Handbook," "The Practical Engineer's Handbook," &c. Fcap. 8vo, nearly 500 pp., with Eight Plates and upwards of 250 Illustrative Diagrams, 6s., strongly bound for workshop or pocket wear and tear. [*Just published.*

*** OPINIONS OF THE PRESS.

"In its modernised form Hutton's 'Templeton' should have a wide sale, for it contains much valuable information which the mechanic will often find of use, and not a few tables and notes which he might look for in vain in other works. This modernised edition wi be appreciated by all who have learned to value the original editions of 'Templeton.'"—*English Mechanic.*

"It has met with great success in the engineering workshop, as we can testify; and there are a great many men who, in a great measure, owe their rise in life to this little book."—*Building News.*

"This familiar text-book—well known to all mechanics and engineers—is of essential service to the every-day requirements of engineers, millwrights, and the various trades connected with engineering and building. The new modernised edition is worth its weight in gold."—*Building News.* (Second Notice.)

"This well-known and largely used book contains information, brought up to date, of the sort so useful to the foreman and draughtsman. So much fresh information has been introduced as to constitute it practically a new book. It will be largely used in the office and workshop. '—*Mechanical World.*

Stone-working Machinery.

STONE-WORKING MACHINERY, and the Rapid and Economical Conversion of Stone. With Hints on the Arrangement and Management of Stone Works. By M. Powis Bale, M.I.M.E. With Illusts. Crown 8vo, 9s.

" Should be in the hands of every mason or student of stone-work."—*Colliery Guardian.*
" A capital handbook for all who manipulate stone for building or ornamental purposes."—*Machinery Market.*

Pump Construction and Management.

PUMPS AND PUMPING: A Handbook for Pump Users. Being Notes on Selection, Construction and Management. By M. Powis Bale, M.I.M.E., Author of " Woodworking Machinery," &c. Crown 8vo, 2s. 6d.

"The matter is set forth as concisely as possible. In fact, condensation rather than diffuseness has been the author's aim throughout; yet he does not seem to have omitted anything likely to be of use."—*Journal of Gas Lighting.*

Milling Machines, etc.

MILLING: A Treatise on Machines, Appliances, and Processes employed in the Shaping of Metals by Rotary Cutters, including Information on Making and Grinding the Cutters. By Paul N. Hasluck, Author of " Lathework." With upwards of 300 Engravings. Large crown 8vo, 12s. 6d. cloth.
[*Just published.*]

Turning.

LATHE-WORK: A Practical Treatise on the Tools, Appliances, and Processes employed in the Art of Turning. By Paul N. Hasluck. Fourth Edition, Revised and Enlarged. Cr. 8vo, 5s. cloth.

" Written by a man who knows, not only how work ought to be done, but who also knows how to do it, and how to convey his knowledge to others. To all turners this book would be valuable."—*Engineering.*
" We can safely recommend the work to young engineers. To the amateur it will simply be invaluable. To the student it will convey a great deal of useful information."—*Engineer.*

Screw-Cutting.

SCREW THREADS: And Methods of Producing Them. With Numerous Tables, and complete directions for using Screw-Cutting Lathes. By Paul N. Hasluck, Author of " Lathe-Work," &c. With Fifty Illustrations. Third Edition, Enlarged. Waistcoat-pocket size, 1s. 6d. cloth.

" Full of useful information, hints and practical criticism. Taps, dies and screwing-tools generally are illustrated and their action described."—*Mechanical World.*
" It is a complete compendium of all the details of the screw cutting lathe; in fact a *multum-in-parvo* on all the subjects it treats upon."—*Carpenter and Builder.*

Smith's Tables for Mechanics, etc.

TABLES, MEMORANDA, AND CALCULATED RESULTS, FOR MECHANICS, ENGINEERS, ARCHITECTS, BUILDERS, etc. Selected and Arranged by Francis Smith. Fifth Edition, thoroughly Revised and Enlarged, with a New Section of Electrical Tables, Formulæ, and Memoranda. Waistcoat-pocket size, 1s. 6d. limp leather. [*Just published.*]

" It would, perhaps, be as difficult to make a small pocket-book selection of notes and formulæ to suit ALL engineers as it would be to make a universal medicine; but Mr. Smith's waistcoat-pocket collection may be looked upon as a successful attempt."—*Engineer.*
" The best example we have ever seen of 250 pages of useful matter packed into the dimensions of a card-case."—*Building News.* " A veritable pocket treasury of knowledge."—*Iron.*

Engineer's and Machinist's Assistant.

THE ENGINEER'S, MILLWRIGHT'S, and MACHINIST'S PRACTICAL ASSISTANT. A collection of Useful Tables, Rules and Data By William Templeton. 7th Edition, with Additions. 18mo, 2s. 6d. cloth.

" Occupies a foremost place among books of this kind. A more suitable present to an apprentice to any of the mechanical trades could not possibly be made."—*Building News.*
" A deservedly popular work, it should be in the 'drawer' of every mechanic."—*English Mechanic.*

Iron and Steel.

"IRON AND STEEL": A Work for the Forge, Foundry, Factory, and Office. Containing ready, useful, and trustworthy Information for Ironmasters; Managers of Bar, Rail, Plate, and Sheet Rolling Mills; Iron and Metal Founders; Iron Ship and Bridge Builders; Mechanical, Mining, and Consulting Engineers; Contractors, Builders, &c. By Charles Hoare. Eighth Edition, Revised and considerably Enlarged. 32mo, 6s. leather.

" One of the best of the pocket books."—*English Mechanic.*
" We cordially recommend this book to those engaged in considering the details of all kinds of Iron and steel works."—*Naval Science.*

Engineering Construction.

PATTERN-MAKING : A Practical Treatise, embracing the Main Types of Engineering Construction, and including Gearing, both Hand and Machine made, Engine Work, Sheaves and Pulleys, Pipes and Columns, Screws, Machine Parts, Pumps and Cocks, the Moulding of Patterns in Loam and Greensand, &c., together with the methods of Estimating the weight of Castings; to which is added an Appendix of Tables for Workshop Reference. By a FOREMAN PATTERN MAKER. With upwards of Three Hundred and Seventy Illustrations. Crown 8vo, 7s. 6d. cloth.

"A well-written technical guide, evidently written by a man who understands and has practised what he has written about. . . . We cordially recommend it to engineering students, young journeymen, and others desirous of being initiated into the mysteries of pattern-making."—*Builder.*
"We can confidently recommend this comprehensive treatise.'—*Building News.*
"Likely to prove a welcome guide to many workmen, especially to draughtsmen who have lacked a training in the shops, pupils pursuing their practical studies in our factories, and to employers and managers in engineering works."—*Hardware Trade Journal.*
"More than 370 illustrations help to explain the text, which is, however, always clear and explicit, thus rendering the work an excellent *vade mecum* for the apprentice who desires to become master of his trade."—*English Mechanic.*

Dictionary of Mechanical Engineering Terms.

LOCKWOOD'S DICTIONARY OF TERMS USED IN THE PRACTICE OF MECHANICAL ENGINEERING, embracing those current in the Drawing Office, Pattern Shop, Foundry, Fitting, Turning, Smith's and Boiler Shops, &c. &c. Comprising upwards of 6,000 Definitions. Edited by A FOREMAN PATTERN-MAKER, Author of "Pattern Making." Crown 8vo, 7s. 6d. cloth.

"Just the sort of handy dictionary required by the various trades engaged in mechanical engineering. The practical engineering pupil will find the book of great value in his studies, and every foreman engineer and mechanic should have a copy."—*Building News.*
"After a careful examination of the book, and trying all manner of words, we think that the engineer will here find all he is likely to require. It will be largely used."—*Practical Engineer.*
"One of the most useful books which can be presented to a mechanic or student."—*English Mechanic.*
"Not merely a dictionary, but, to a certain extent, also a most valuable guide. It strikes us as a happy idea to combine with a definition of the phrase useful information on the subject of which it treats."—*Machinery Market.*
"No word having connection with any branch of constructive engineering seems to be omitted. No more comprehensive work has been, so far, issued. —*Knowledge.*
"We strongly commend this useful and reliable adviser to our friends in the workshop, and to students everywhere."—*Colliery Guardian.*

Steam Boilers.

A TREATISE ON STEAM BOILERS: Their Strength, Construction, and Economical Working. By ROBERT WILSON, C.E. Fifth Edition. 12mo, 6s. cloth.

"The best treatise that has ever been published on steam boilers."—*Engineer.*
"The author shows himself perfect master of his subject, and we heartily recommend all employing steam power to possess themselves of the work."—*Ryland's Iron Trade Circular.*

Boiler Chimneys.

BOILER AND FACTORY CHIMNEYS: Their Draught-Power and Stability. With a Chapter on *Lightning Conductors.* By ROBERT WILSON, A.I.C.E., Author of "A Treatise on Steam Boilers," &c. Second Edition. Crown 8vo, 3s. 6d. cloth.

"Full of useful information, definite in statement, and thoroughly practical in treatment. — *The Local Government Chronicle.*
"A valuable contribution to the iterature of scientific building."—*The Builder.*

Boiler Making.

THE BOILER-MAKER'S READY RECKONER & ASSISTANT. With Examples of Practical Geometry and Templating, for the Use of Platers, Smiths and Riveters. By JOHN COURTNEY. Edited by D. K. CLARK, M.I.C.E. Third Edition, 480 pp., with 140 Illusts. Fcap. 8vo, 7s. half-bound.

"No workman or apprentice should be without this book."—*Iron Trade Circular.*
"Boiler-makers will readily recognise the value of this volume. . . . The tables are clearly printed, and so arranged that they can be referred to with the greatest facility, so that it cannot be doubted that they will be generally appreciated and much used."—*Mining Journal.*

Warming.

HEATING BY HOT WATER; with Information and Suggestions on the best Methods of Heating Public, Private and Horticultural Buildings. By WALTER JONES. With Illustrations, crown 8vo, 2s. cloth.

"We confidently recommend all interested in heating by hot water to secure a copy of this valuable little treatise."—*The Plumber and Decorator.*

Steam Engine.

TEXT-BOOK ON THE STEAM ENGINE. With a Supplement on Gas Engines, and PART II. ON HEAT ENGINES. By T. M. GOODEVE, M.A., Barrister-at-Law, Professor of Mechanics at the Normal School of Science and the Royal School of Mines; Author of "The Principles of Mechanics," "The Elements of Mechanism," &c. Eleventh Edition, Enlarged. With numerous Illustrations. Crown 8vo, 6s. cloth.

"Professor Goodeve has given us a treatise on the steam engine which will bear comparison with anything written by Huxley or Maxwell, and we can award it no higher praise."—*Engineer*

" Mr. Goodeve's text-book is a work of which every young engineer should possess himself."—*Mining Journal.*

Gas Engines.

ON GAS-ENGINES. Being a Reprint, with some Additions, of the Supplement to the *Text-book on the Steam Engine,* by T. M. GOODEVE, M.A. Crown 8vo, 2s. 6d. cloth.

" Like all Mr. Goodeve's writings, the present is no exception in point of general excellence. It is a valuable little volume."—*Mechanical World.*

Steam.

THE SAFE USE OF STEAM. Containing Rules for Unprofessional Steam-users. By an ENGINEER. Sixth Edition. Sewed, 6d.

"If steam-users would but learn this little book by heart boiler explosions would become sensations by their rarity."—*English Mechanic.*

Reference Book for Mechanical Engineers.

THE MECHANICAL ENGINEER'S REFERENCE BOOK, for Machine and Boiler Construction. In Two Parts. Part I. GENERAL ENGINEERING DATA. Part II. BOILER CONSTRUCTION. With 51 Plates and numerous Illustrations. By NELSON FOLEY, M.I.N.A. Folio, £5 5s. half-bound. [*Just published.*

Coal and Speed Tables.

A POCKET BOOK OF COAL AND SPEED TABLES, *for Engineers and Steam-users.* By NELSON FOLEY, Author of "Boiler Construction." Pocket-size, 3s. 6d. cloth; 4s. leather.

"These tables are designed to meet the requirements of every-day use ; and may be commended to engineers and users of steam."—*Iron.*

"This pocket-book well merits the attention of the practical engineer. Mr. Foley has compiled a very useful set of tables, the information contained in which is frequently required by engineers, coal consumers and users of steam."—*Iron and Coal Trades Review.*

Fire Engineering.

FIRES, FIRE-ENGINES, AND FIRE-BRIGADES. With a History of Fire-Engines, their Construction, Use, and Management ; Remarks on Fire-Proof Buildings, and the Preservation of Life from Fire ; Foreign Fire Systems, &c. By C. F. T. YOUNG, C.E. With numerous Illustrations, 544 pp., demy 8vo, £1 4s. cloth.

" To such of our readers as are interested in the subject of fires and fire apparatus, we can in heartily commend this book."—*Engineering.*

"It displays much evidence of careful research ; and Mr. Young has put his facts neatly together. It is evident enough that his acquaintance with the practical details of the construction steam fire engines is accurate and full."—*Engineer.*

Estimating for Engineering Work, &c.

ENGINEERING ESTIMATES, COSTS AND ACCOUNTS : A Guide to Commercial Engineering. With numerous Examples of Estimates and Costs of Millwright Work, Miscellaneous Productions, Steam Engines and Steam Boilers; and a Section on the Preparation of Costs Accounts. By A GENERAL MANAGER. Demy 8vo, 12s. cloth.

" This is an excellent and very useful book, covering subject-matter in constant requisition in every factory and workshop. . . . The book is invaluable, not only to the young engineer, but also to the estimate department of every works."—*Builder.*

" We accord the work unqualified praise. The information is given in a plain, straightforward manner and bears throughout evidence of the intimate practical acquaintance of the author with every phrase of commercial engineering."—*Mechanical World.*

Elementary Mechanics.

CONDENSED MECHANICS. A Selection of Formulæ, Rules, Tables, and Data for the Use of Engineering Students, Science Classes, &c. In Accordance with the Requirements of the Science and Art Department. By W. G. CRAWFORD HUGHES, A.M.I.C.E. Crown 8vo, 2s. 6d. cloth.

[*Just published.*

THE POPULAR WORKS OF MICHAEL REYNOLDS
("The Engine Driver's Friend").

Locomotive-Engine Driving.

LOCOMOTIVE-ENGINE DRIVING : A Practical Manual for Engineers in charge of Locomotive Engines. By Michael Reynolds, Member of the Society of Engineers, formerly Locomotive Inspector L. B. and S. C. R. Eighth Edition. Including a Key to the Locomotive Engine. With Illustrations and Portrait of Author. Crown 8vo, 4s. 6d. cloth.

"Mr. Reynolds has supplied a want, and has supplied it well. We can confidently recommend the book, not only to the practical driver, but to everyone who takes an interest in the performance of locomotive engines."—The Engineer.

"Mr. Reynolds has opened a new chapter in the literature of the day. This admirable practical treatise, of the practical utility of which we have to speak in terms of warm commendation."—Athenaeum.

"Evidently the work of one who knows his subject thoroughly."—Railway Service Gazette.

"Were the cautions and rules given in the book to become part of the every-day working of our engine-drivers, we might have fewer distressing accidents to deplore."—Scotsman.

Stationary Engine Driving.

STATIONARY ENGINE DRIVING : A Practical Manual for Engineers in charge of Stationary Engines. By Michael Reynolds. Fourth Edition, Enlarged. With Plates and Woodcuts. Crown 8vo, 4s. 6d. cloth.

"The author is thoroughly acquainted with his subjects, and his advice on the various points treated is clear and practical. . . . He has produced a manual which is an exceedingly useful one for the class for whom it is specially intended."—Engineering.

"Our author leaves no stone unturned. He is determined that his readers shall not only know something about the stationary engine, but all about it."—Engineer.

"An engineman who has mastered the contents of Mr. Reynolds's book will require but little actual experience with boilers and engines before he can be trusted to look after them."—English Mechanic.

The Engineer, Fireman, and Engine-Boy.

THE MODEL LOCOMOTIVE ENGINEER, FIREMAN, and ENGINE-BOY. Comprising a Historical Notice of the Pioneer Locomotive Engines and their Inventors. By Michael Reynolds. With numerous Illustrations and a fine Portrait of George Stephenson. Crown 8vo, 4s. 6d. cloth.

"From the technical knowledge of the author it will appeal to the railway man of to-day more forcibly than anything written by Dr. Smiles. . . . The volume contains information of a technical kind, and facts that every driver should be familiar with."—English Mechanic.

"We should be glad to see this book in the possession of everyone in the kingdom who has ever laid, or is to lay, hands on a locomotive engine."—Iron.

Continuous Railway Brakes.

CONTINUOUS RAILWAY BRAKES : A Practical Treatise on the several Systems in Use in the United Kingdom ; their Construction and Performance. With copious Illustrations and numerous Tables. By Michael Reynolds. Large crown 8vo, 9s. cloth.

"A popular explanation of the different brakes. It will be of great assistance in forming public opinion, and will be studied with benefit by those who take an interest in the brake."—English Mechanic.

"Written with sufficient technical detail to enable the principle and relative connection of the various parts of each particular brake to be readily grasped."—Mechanical World.

Engine-Driving Life.

ENGINE-DRIVING LIFE : Stirring Adventures and Incidents in the Lives of Locomotive-Engine Drivers. By Michael Reynolds. Second Edition, with Additional Chapters. Crown 8vo, 2s. cloth.

"From first to last perfectly fascinating. Wilkie Collins's most thrilling conceptions are thrown into the shade by true incidents, endless in their variety, related in every page."—North British Mail.

"Anyone who wishes to get a real insight into railway life cannot do better than read 'Engine-Driving Life' for himself ; and if he once take it up he will find that the author's enthusiasm and real love of the engine-driving profession will carry him on till he has read every page."—Saturday Review.

Pocket Companion for Enginemen.

THE ENGINEMAN'S POCKET COMPANION AND PRACTICAL EDUCATOR FOR ENGINEMEN, BOILER ATTENDANTS, AND MECHANICS. By Michael Reynolds. With Forty-five Illustrations and numerous Diagrams. Second Edition, Revised. Royal 18mo, 3s. 6d., strongly bound for pocket wear.

"This admirable work is well suited to accomplish its object, being the honest workmanship of a competent engineer."—Glasgow Herald.

"A most meritorious work, giving in a succinct and practical form all the information an engine-minder desirous of mastering the scientific principles of his daily calling would require."—Miller.

"A boon to those who are striving to become efficient mechanics."—Daily Chronicle.

French-English Glossary for Engineers, etc.

A POCKET GLOSSARY of TECHNICAL TERMS : ENGLISH-FRENCH, FRENCH-ENGLISH ; with Tables suitable for the Architectural, Engineering, Manufacturing and Nautical Professions. By JOHN JAMES FLETCHER, Engineer and Surveyor. 200 pp. Waistcoat-pocket size, 1s. 6d., limp leather.

"It ought certainly to be in the waistcoat-pocket of every professional man."—*Iron.*

"It is a very great advantage for readers and correspondents in France and England to have so large a number of the words relating to engineering and manufacturers collected in a lilliputian volume. The little book will be useful both to students and travellers.'—*Architect.*

"The glossary of terms is very complete, and many of the tables are new and well arranged. We cordially commend the book."—*Mechanical World*

Portable Engines.

THE PORTABLE ENGINE; ITS CONSTRUCTION AND MANAGEMENT. A Practical Manual for Owners and Users of Steam Engines generally. By WILLIAM DYSON WANSBROUGH. With 90 Illustrations. Crown 8vo, 3s. 6d. cloth.

"This is a work of value to those who use steam machinery. . . . Should be read by everyone who has a steam engine, on a farm or elsewhere."—*Mark Lane Express.*

"We cordially commend this work to buyers and owners of steam engines, and to those who have to do with their construction or use."—*Timber Trades Journal.*

"Such a general knowledge of the steam engine as Mr. Wansbrough furnishes to the reader should be acquired by all intelligent owners and others who use the steam engine."—*Building News.*

"An excellent text-book of this useful form of engine, which describes with all necessary minuteness the details of the various devices. . . The Hints to Purchasers contain a good deal of commonsense and practical wisdom."—*English Mechanic.*

CIVIL ENGINEERING, SURVEYING, etc.

MR. HUMBER'S IMPORTANT ENGINEERING BOOKS.

The Water Supply of Cities and Towns.

A COMPREHENSIVE TREATISE on the WATER-SUPPLY OF CITIES AND TOWNS. By WILLIAM HUMBER, A-M.Inst.C.E., and M. Inst. M.E., Author of "Cast and Wrought Iron Bridge Construction," &c. &c. Illustrated with 50 Double Plates, 1 Single Plate, Coloured Frontispiece, and upwards of 250 Woodcuts, and containing 400 pages of Text. Imp. 4to, £6 6s. elegantly and substantially half-bound in morocco.

List of Contents.

I. Historical Sketch of some of the means that have been adopted for the Supply of Water to Cities and Towns.—II. Water and the Foreign Matter usually associated with it.—III. Rainfall and Evaporation.—IV. Springs and the water-bearing formations of various districts.—V. Measurement and Estimation of the flow of Water —VI. On the Selection of the Source of Supply.—VII. Wells.—VIII. Reservoirs.—IX. The Purification of Water.—X. Pumps. — XI. Pumping Machinery. — XII. Conduits.—XIII. Distribution of Water.—XIV. Meters, Service Pipes, and House Fittings.—XV. The Law and Economy of Water Works.—XVI. Constant and Intermittent Supply.—XVII. Description of Plates. — Appendices, giving Tables of Rates of Supply, Velocities, &c. &c., together with Specifications of several Works illustrated, among which will be found : Aberdeen, Bideford, Canterbury, Dundee, Halifax, Lambeth, Rotherham, Dublin, and others.

"The most systematic and valuable work upon water supply hitherto produced in English, or in any other language. . . . Mr. Humber's work is characterised almost throughout by an exhaustiveness much more distinctive of French and German than of English technical treatises."—*Engineer.*

"We can congratulate Mr. Humber on having been able to give so large an amount of information on a subject so important as the water supply of cities and towns. The plates, fifty in number, are mostly drawings of executed works, and alone would have commanded the attention of every engineer whose practice may lie in this branch of the profession."—*Builder.*

Cast and Wrought Iron Bridge Construction.

A COMPLETE AND PRACTICAL TREATISE ON CAST AND WROUGHT IRON BRIDGE CONSTRUCTION, including Iron Foundations. In Three Parts—Theoretical, Practical, and Descriptive. By WILLIAM HUMBER, A.M.Inst.C.E., and M.Inst.M.E. Third Edition, Revised and much improved, with 115 Double Plates (20 of which now first appear in this edition), and numerous Additions to the Text. In Two Vols., imp. 4to, £6 16s. 6d. half-bound in morocco.

"A very valuable contribution to the standard literature of civil engineering. In addition to elevations, plans and sections, large scale details are given which very much enhance the instructive worth of those illustrations."—*Civil Engineer and Architect's Journal.*

"Mr. Humber's stately volumes, lately issued—in which the most important bridges erected during the last five years, under the direction of the late Mr. Brunel, Sir W. Cubitt, Mr. Hawkshaw, Mr. Page, Mr. Fowler, Mr. Hemans, and others among our most eminent engineers, are drawn and specified in great detail."—*Engineer.*

MR. HUMBER'S GREAT WORK ON MODERN ENGINEERING.

Complete in Four Volumes, imperial 4to, price £12 12s., half-morocco. Each
Volume sold separately as follows:—

A RECORD OF THE PROGRESS OF MODERN ENGINEER-
ING. FIRST SERIES. Comprising Civil, Mechanical, Marine, Hydraulic,
Railway, Bridge, and other Engineering Works, &c. By WILLIAM HUMBER,
A-M.Inst.C.E., &c. Imp. 4to, with 36 Double Plates, drawn to a large scale,
Photographic Portrait of John Hawkshaw, C.E., F.R.S., &c., and copious
descriptive Letterpress, Specifications, &c., £3 3s. half-morocco.

List of the Plates and Diagrams.

Victoria Station and Roof, L. B. & S. C. R. (8 plates); Southport Pier (2 plates); Victoria Station and Roof, L. C. & D. and G. W. R. (6 plates); Roof of Cremorne Music Hall; Bridge over G. N. Railway; Roof of Station, Dutch Rhenish Rail (2 plates); Bridge over the Thames, West London Extension Railway (5 plates); Armour Plates: Suspension Bridge, Thames (4 plates); The Allen Engine; Suspension Bridge, Avon (3 plates); Underground Railway (3 plates).

"Handsomely lithographed and printed. It will find favour with many who desire to preserve in a permanent form copies of the plans and specifications prepared for the guidance of the contractors for many important engineering works."—*Engineer.*

HUMBER'S RECORD OF MODERN ENGINEERING. SECOND
SERIES. Imp. 4to, with 36 Double Plates, Photographic Portrait of Robert
Stephenson, C.E., M.P., F.R.S., &c., and copious descriptive Letterpress,
Specifications, &c., £3 3s. half-morocco.

List of the Plates and Diagrams.

Birkenhead Docks, Low Water Basin (15 plates); Charing Cross Station Roof, C. C. Railway (3 plates); Digswell Viaduct, Great Northern Railway; Robbery Wood Viaduct, Great Northern Railway; Iron Permanent Way; Clydach Viaduct, Merthyr, Tredegar, and Abergavenny Railway; Ebbw Vladuct, Merthyr, Tredegar, and Ahergavenny Railway; College Wood Viaduct, Cornwall Railway; Dublin Winter Palace Roof (3 plates); Bridge over the Thames, L. C. & D. Railway (6 plates); Albert Harbour, Greenock (4 plates).

"Mr. Humber has done the profession good and true service, by the fine collection of examples he has here brought before the profession and the public."—*Practical Mechanic's Journal.*

HUMBER'S RECORD OF MODERN ENGINEERING. THIRD
SERIES. Imp. 4to, with 40 Double Plates, Photographic Portrait of J. R.
M'Clean, late Pres. Inst. C.E., and copious descriptive Letterpress, Speci-
fications, &c., £3 3s. half-morocco.

List of the Plates and Diagrams.

MAIN DRAINAGE, METROPOLIS.—*North Side.*—Map showing Interception of Sewers; Middle Level Sewer (2 plates); Outfall Sewer, Bridge over River Lea (3 plates); Outfall Sewer, Bridge over Marsh Lane, North Woolwich Railway, and Bow and Barking Railway Junction; Outfall Sewer, Bridge over Bow and Barking Railway (3 plates); Outfall Sewer, Bridge over East London Waterworks' Feeder (2 plates); Outfall Sewer, Reservoir (2 plates); Outfall Sewer, Tumbling Bay and Outlet; Outfall Sewer, Penstocks. *South Side.*—Outfall Sewer, Bermondsey Branch (2 plates); Outfall Sewer, Reservoir and Outlet (4 plates); Outfall Sewer, Filth Hoist; Sections of Sewers (North and South Sides).
THAMES EMBANKMENT.—Section of River Wall; Steamboat Pier, Westminster (2 plates); Landing Stairs between Charing Cross and Waterloo Bridges; York Gate (2 plates); Overflow and Outlet at Savoy Street Sewer (3 plates); Steamboat Pier, Waterloo Bridge (3 plates); Junction of Sewers, Plans and Sections; Gullies, Flans and Sections; Rolling Stock; Granite and Iron Forts.

"The drawings have a constantly increasing value, and whoever desires to possess clear representations of the two great works carried out by our Metropolitan Board will obtain Mr. Humber's volume."—*Engineer.*

HUMBER'S RECORD OF MODERN ENGINEERING. FOURTH
SERIES. Imp. 4to, with 36 Double Plates, Photographic Portrait of John
Fowler, late Pres. Inst. C.E., and copious descriptive Letterpress, Speci-
fications, &c., £3 3s. half-morocco.

List of the Plates and Diagrams.

Abbey Mills Pumping Station, Main Drainage, Metropolis (4 plates); Barrow Docks (5 plates); Manquis Viaduct, Santiago and Valparaiso Railway (2 plates); Adam's Locomotive, St. Helen's Canal Railway (2 plates); Cannon Street Station Roof, Charing Cross Railway (3 plates); Road Bridge over the River Moka (2 plates); Telegraphic Apparatus for Mesopotamia; Viaduct over the River Wye, Midland Railway (3 plates); St. Germans Viaduct, Cornwall Railway (2 plates); Wrought-Iron Cylinder for Diving Bell; Millwall Docks (6 plates); Milroy's Patent Excavator; Metropolitan District Railway (6 plates); Harbours, Ports, and Breakwaters (3 plates).

"We gladly welcome another year's issue of this valuable publication from the able pen of Mr. Humber. The accuracy and general excellence of this work are well known, while its usefulness in giving the measurements and details of some of the latest examples of engineering, as carried out by the most eminent men in the profession, cannot be too highly prized."—*Artisan.*

Strains, Calculation of.

A HANDY BOOK FOR THE CALCULATION OF STRAINS IN GIRDERS AND SIMILAR STRUCTURES, AND THEIR STRENGTH. Consisting of Formulæ and Corresponding Diagrams, with numerous details for Practical Application, &c. By WILLIAM HUMBER, A·M.Inst.C.E., &c. Fourth Edition. Crown 8vo, nearly 100 Woodcuts and 3 Plates, 7s. 6d. cloth.

" The formulæ are neatly expressed, and the diagrams good."—*Athenæum.*
" We heartily commend this really *handy* book to our engineer and architect readers."—*English Mechanic.*

Barlow's Strength of Materials, enlarged by Humber

A TREATISE ON THE STRENGTH OF MATERIALS; with Rules for Application in Architecture, the Construction of Suspension Bridges, Railways, &c. By PETER BARLOW, F.R.S. A New Edition, revised by his Sons, P. W. BARLOW, F.R.S., and W. H. BARLOW, F.R.S.; to which are added, Experiments by HODGKINSON, FAIRBAIRN, and KIRKALDY; and Formulæ for Calculating Girders, &c. Arranged and Edited by W. HUMBER, A·M.Inst.C.E. Demy 8vo, 400 pp., with 19 large Plates and numerous Wood-cuts, 18s. cloth.

" Valuable alike to the student, tyro, and the experienced practitioner, it will always rank in future, as it has hitherto done, as the standard treatise on that particular subject."—*Engineer.*
" There is no greater authority than Barlow."—*Building News.*
" As a scientific work of the first class. it deserves a foremost place on the bookshelves of every civil engineer and practical mechanic."—*English Mechanic.*

Trigonometrical Surveying.

AN OUTLINE OF THE METHOD OF CONDUCTING A TRIGONOMETRICAL SURVEY, for the Formation of Geographical and Topographical Maps and Plans, Military Reconnaissance, Levelling, &c., with Useful Problems, Formulæ, and Tables. By Lieut.-General FROME, R.E. Fourth Edition, Revised and partly Re-written by Major General Sir CHARLES WARREN, G.C.M.G., R.E. With 19 Plates and 115 Woodcuts, royal 8vo, 16s. cloth.

" The simple fact that a fourth edition has been called for is the best testimony to its merits. No words of praise from us can strengthen the position so well and so steadily maintained by this work. Sir Charles Warren has revised the entire work, and made such additions as were necessary to bring every portion of the contents up to the present date."—*Broad Arrow.*

Field Fortification.

A TREATISE ON FIELD FORTIFICATION, THE ATTACK OF FORTRESSES, MILITARY MINING, AND RECONNOITRING. By Colonel I. S. MACAULAY, late Professor of Fortification in the R.M.A., Woolwich. Sixth Edition, crown 8vo, cloth, with separate Atlas of 12 Plates, 12s.

Oblique Bridges.

A PRACTICAL AND THEORETICAL ESSAY ON OBLIQUE BRIDGES. With 13 large Plates. By the late GEORGE WATSON BUCK, M.I.C.E. Third Edition, revised by his Son, J. H. WATSON BUCK, M.I.C.E.; and with the addition of Description to Diagrams for Facilitating the Construction of Oblique Bridges, by W. H. BARLOW, M.I.C.E. Royal 8vo, 12s. cloth.

" The standard text-book for all engineers regarding skew arches is Mr. Buck's treatise, and it would be impossible to consult a better."—*Engineer.*
" Mr. Buck's treatise is recognised as a standard text-book, and his treatment has divested the subject of many of the intricacies supposed to belong to it. As a guide to the engineer and architect, on a confessedly difficult subject, Mr. Buck's work is unsurpassed."—*Building News.*

Water Storage, Conveyance and Utilisation.

WATER ENGINEERING : A Practical Treatise on the Measurement, Storage, Conveyance and Utilisation of Water for the Supply of Towns, for Mill Power, and for other Purposes. By CHARLES SLAGG, Water and Drainage Engineer, A.M.Inst.C.E., Author of "Sanitary Work in the Smaller Towns, and in Villages," &c. With numerous Illusts. Cr. 8vo. 7s. 6d. cloth.

" As a small practical treatise on the water supply of towns, and on some applications of water-power, the work is in many respects excellent."—*Engineering.*
" The author has collated the results deduced from the experiments of the most eminent authorities, and has presented them in a compact and practical form, accompanied by very clear and detailed explanations. . . . The application of water as a motive power is treated very carefully and exhaustively."—*Builder.*
" For anyone who desires to begin the study of hydraulics with a consideration of the practical applications of the science there is no better guide."—*Architect.*

Statics, Graphic and Analytic.

GRAPHIC AND ANALYTIC STATICS, *in their Practical Application to the Treatment of Stresses in Roofs, Solid Girders, Lattice, Bowstring and Suspension Bridges, Braced Iron Arches and Piers, and other Frameworks.* By R. HUDSON GRAHAM, C.E. Containing Diagrams and Plates to Scale. With numerous Examples, many taken from existing Structures. Specially arranged for Class-work in Colleges and Universities. Second Edition, Revised and Enlarged. 8vo, 16s. cloth.

"Mr. Graham's book will find a place wherever graphic and analytic statics are used or studied." —*Engineer.*

"The work is excellent from a practical point of view, and has evidently been prepared with much care. The directions for working are ample, and are illustrated by an abundance of well-selected examples. It is an excellent text-book for the practical draughtsman."—*Athenæum.*

Student's Text-Book on Surveying.

PRACTICAL SURVEYING : A Text-Book for Students preparing for Examination or for Survey-work in the Colonies. By GEORGE W. USILL, A.M.I.C.E., Author of "The Statistics of the Water Supply of Great Britain." With Four Lithographic Plates and upwards of 330 Illustrations. Second Edition, Revised. Crown 8vo, 7s. 6d. cloth.

"The best forms of instruments are described as to their construction, uses and modes of employment, and there are innumerable hints on work and equipment such as the author, in his experience as surveyor, draughtsman and teacher, has found necessary, and which the student in his inexperience will find most serviceable."—*Engineer.*

"The latest treatise in the English language on surveying, and we have no hesitation in saying that the student will find it a better guide than any of its predecessors . . . . Deserves to be recognised as the first book which should be put in the hands of a pupil of Civil Engineering, and every gentleman of education who sets out for the Colonies would find it well to have a copy."—*Architect.*

"A very useful, practical handbook on field practice. Clear, accurate and not too con densed."—*Journal of Education.*

Survey Practice.

AID TO SURVEY PRACTICE, *for Reference in Surveying, Levelling, and Setting-out ; and in Route Surveys of Travellers by Land and Sea.* With Tables, Illustrations, and Records. By LOWIS D'A. JACKSON, A.M.I.C.E., Author of "Hydraulic Manual," "Modern Metrology," &c. Second Edition, Enlarged. Large crown 8vo, 12s. 6d. cloth.

"Mr. Jackson has produced a valuable *vade-mecum* for the surveyor. We can recommend this book as containing an admirable supplement to the teaching of the accomplished surveyor."—*Athenæum.*

"As a text-book we should advise all surveyors to place it in their libraries, and study well the matured instructions afforded in its pages."—*Colliery Guardian.*

"The author brings to his work a fortunate union of theory and practical experience which, aided by a clear and lucid style of writing, renders the book a very useful one."—*Builder.*

Surveying, Land and Marine.

LAND AND MARINE SURVEYING, in Reference to the Preparation of Plans for Roads and Railways; Canals, Rivers, Towns' Water Supplies; Docks and Harbours. With Description and Use of Surveying Instruments. By W. D. HASKOLL, C.E., Author of "Bridge and Viaduct Construction," &c. Second Edition, Revised, with Additions. Large cr. 8vo, 9s. cl.

"This book must prove of great value to the student. We have no hesitation in recommending it, feeling assured that it will more than repay a careful study."—*Mechanical World.*

"A most useful and well arranged book for the aid of a student. We can strongly recommend it as a carefully-written and valuable text-book. It enjoys a well-deserved repute among surveyors." —*Builder.*

"This volume cannot fail to prove of the utmost practical utility. It may be safely recommended to all students who aspire to become clean and expert surveyors."—*Mining Journal.*

Tunnelling.

PRACTICAL TUNNELLING. Explaining in detail the Setting-out of the works, Shaft-sinking and Heading-driving, Ranging the Lines and Levelling underground, Sub-Excavating, Timbering, and the Construction of the Brickwork of Tunnels, with the amount of Labour required for, and the Cost of, the various portions of the work. By FREDERICK W. SIMMS, F.G.S., M.Inst.C.E. Third Edition, Revised and Extended by D. KINNEAR CLARK, M.Inst.C.E. Imperial 8vo, with 21 Folding Plates and numerous Wood Engravings, 30s. cloth.

"The estimation in which Mr. Simms's book on tunnelling has been held for over thirty years cannot be more truly expressed than in the words of the late Prof. Rankine :—' The best source of information on the subject of tunnels is Mr. F. W. Simms's work on Practical Tunnelling.'"—*Architect.*

"It has been regarded from the first as a text book of the subject. . . . Mr. Clarke has added immensely to the value of the book."—*Engineer.*

Levelling.

A TREATISE ON THE PRINCIPLES AND PRACTICE OF LEVELLING. Showing its Application to purposes of Railway and Civil Engineering, in the Construction of Roads; with Mr. TELFORD's Rules for the same. By FREDERICK W. SIMMS, F.G.S., M.Inst.C.E. Seventh Edition, with the addition of LAW's Practical Examples for Setting-out Railway Curves, and TRAUTWINE's Field Practice of Laying-out Circular Curves. With 7 Plates and numerous Woodcuts, 8vo, 8s. 6d. cloth. *.* TRAUTWINE on Curves may be had separate, 5s.

"The text-book on levelling in most of our engineering schools and colleges."—*Engineer.*

"The publishers have rendered a substantial service to the profession, especially to the younger members, by bringing out the present edition of Mr. Simms's useful work."—*Engineering.*

Heat, Expansion by.

EXPANSION OF STRUCTURES BY HEAT. By JOHN KEILY, C.E., late of the Indian Public Works and Victorian Railway Departments. Crown 8vo, 3s. 6d. cloth.

SUMMARY OF CONTENTS.

Section I. FORMULAS AND DATA.	Section VI. MECHANICAL FORCE OF HEAT.
Section II. METAL BARS.	
Section III. SIMPLE FRAMES.	Section VII. WORK OF EXPANSION AND CONTRACTION.
Section IV. COMPLEX FRAMES AND PLATES.	
	Section VIII. SUSPENSION BRIDGES.
Section V. THERMAL CONDUCTIVITY.	Section IX. MASONRY STRUCTURES.

"The aim the author has set before him, viz., to show the effects of heat upon metallic and other structures, is a laudable one, for this is a branch of physics upon which the engineer or architect can find but little reliable and comprehensive data in books."—*Builder.*

"Whoever is concerned to know the effect of changes of temperature on such structures as suspension bridges and the like, could not do better than consult Mr. Keily's valuable and handy exposition of the geometrical principles involved in these changes."—*Scotsman.*

Practical Mathematics.

MATHEMATICS FOR PRACTICAL MEN: Being a Commonplace Book of Pure and Mixed Mathematics. Designed chiefly for the use of Civil Engineers, Architects and Surveyors. By OLINTHUS GREGORY, LL.D., F.R.A.S., Enlarged by HENRY LAW, C.E. 4th Edition, carefully Revised by J. R. YOUNG, formerly Professor of Mathematics, Belfast College. With 13 Plates, 8vo, £1 1s. cloth.

"The engineer or architect will here find ready to his hand rules for solving nearly every mathematical difficulty that may arise in his practice. The rules are in all cases explained by means of examples, in which every step of the process is clearly worked out."—*Builder.*

"One of the most serviceable books for practical mechanics. . . It is an instructive book for the student, and a text-book for him who, having once mastered the subjects it treats of, needs occasionally to refresh his memory upon them."—*Building News.*

Hydraulic Tables.

HYDRAULIC TABLES, CO-EFFICIENTS, and FORMULÆ for finding the Discharge of Water from Orifices, Notches, Weirs, Pipes, and Rivers. With New Formulæ, Tables, and General Information on Rainfall, Catchment-Basins, Drainage, Sewerage, Water Supply for Towns and Mill Power. By JOHN NEVILLE, Civil Engineer, M.R.I.A. Third Ed., carefully Revised, with considerable Additions. Numerous Illusts. Cr. 8vo, 14s. cloth.

"Alike valuable to students and engineers in practice; its study will prevent the annoyance of avoidable failures, and assist them to select the readiest means of successfully carrying out any given work connected with hydraulic engineering."—*Mining Journal.*

"It is, of all English books on the subject, the one nearest to completeness. . . . From the good arrangement of the matter, the clear explanations, and abundance of formulæ, the carefully calculated tables, and, above all, the thorough acquaintance with both theory and construction, which is displayed from first to last, the book w be found to be an acquisition."—*Architect.*

Hydraulics.

HYDRAULIC MANUAL. Consisting of Working Tables and Explanatory Text. Intended as a Guide in Hydraulic Calculations and Field Operations. By LOWIS D'A. JACKSON, Author of "Aid to Survey Practice," "Modern Metrology," &c. Fourth Edition, Enlarged. Large cr. 8vo, 16s. cl.

"The author has had a wide experience in hydraulic engineering and has been a careful observer of the facts which have come under his notice, and from the great mass of material at his command he has constructed a manual which may be accepted as a trustworthy guide to this branch of the engineer's profession. We can heartily recommend this volume to all who desire to be acquainted with the latest development of this important subject." *Engineering.*

"The standard-work in this department of mechanics."—*Scotsman.*

"The most useful feature of this work is its freedom from what is superannuated, and its thorough adoption of recent experiments; the text is, in fact, in great part a short account of the great modern experiments."—*Nature.*

Drainage.

ON THE DRAINAGE OF LANDS, TOWNS AND BUILD
INGS. By G. D. DEMPSEY, C.E., Author of "The Practical Railway En
gineer," &c. Revised, with large Additions on ,RECENT PRACTICE IN
DRAINAGE ENGINEERING, by D. KINNEAR CLARK, M.Inst.C.E. Author of
"Tramways: Their Construction and Working," "A Manual of Rules, Tables,
and Data for Mechanical Engineers." &c. &c. Crown 8vo, 7s. 6d. cloth.

"The new matter added to Mr. Dempsey's excellent work is characterised by the comprehensive grasp and accuracy of detail for which the name of Mr. D. K. Clark is a sufficient voucher."—
Athenæum.

"As a work on recent practice in drainage engineering, the book is to be commended to all
who are making that branch of engineering science their special study."—*Iron.*

"A comprehensive manual on drainage engineering, and a useful introduction to the student."
Building News.

Tramways and their Working.

TRAMWAYS: THEIR CONSTRUCTION AND WORKING
Embracing a Comprehensive History of the System; with an exhaustive
Analysis of the various Modes of Traction, including Horse-Power, Steam,
Heated Water, and Compressed Air; a Description of the Varieties of Rolling
Stock; and ample Details of Cost and Working Expenses: the Progress
recently made in Tramway Construction, &c. &c. By D. KINNEAR CLARK,
M.Inst.C.E. With over 200 Wood Engravings, and 13 Folding Plates. Two
Vols., large crown 8vo, 30s. cloth.

"All interested in tramways must refer to it, as all railway engineers have turned to the author's
work 'Railway Machinery.'"—*Engineer.*

"An exhaustive and practical work on tramways, in which the history of this kind of locomotion, and a description and cost of the various modes of laying tramways, are to be found."—
Building News.

"The best form of rails, the best mode of construction, and the best mechanical appliances
are so fairly indicated in the work under review, that any engineer about to construct a tramway
will be enabled at once to obtain the practical information which will be of most service to him."—
Athenæum.

Oblique Arches.

A PRACTICAL TREATISE ON THE CONSTRUCTION OF
OBLIQUE ARCHES. By JOHN HART. Third Edition, with Plates. Imperial 8vo, 8s. cloth.

Curves, Tables for Setting-out.

TABLES OF TANGENTIAL ANGLES AND MULTIPLES
for Setting-out Curves from 5 to 200 Radius. By ALEXANDER BEAZELEY,
M.Inst.C.E. Third Edition. Printed on 48 Cards, and sold in a cloth box,
waistcoat-pocket size, 3s. 6d.

"Each table is printed on a small card, which, being placed on the theodolite, leaves the hands
free to manipulate the instrument—no small advantage as regards the rapidity of work."—*Engineer.*

"Very handy; a man may know that all his day's work must fa on two of these cards, which
he puts into his own card-case, and leaves the rest behind."—*Athenæum.*

Earthwork.

EARTHWORK TABLES. Showing the Contents in Cubic
Yards of Embankments, Cuttings, &c., of Heights or Depths up to an average
of 80 feet. By JOSEPH BROADBENT, C.E., and FRANCIS CAMPIN, C.E. Crown
8vo, 5s. cloth.

"The way in which accuracy is attained, by a simple division of each cross section into three
el ments, two in which are constant and one variable, is ingenious."—*Athenæum.*

Tunnel Shafts.

THE CONSTRUCTION OF LARGE TUNNEL SHAFTS: A
Practical and Theoretical Essay. By J. H. WATSON BUCK, M.Inst.C.E.,
Resident Engineer, London and North-Western Railway. Illustrated with
Folding Plates, royal 8vo, 12s. cloth.

"Many of the methods given are of extreme practical value to the mason; and the observations
on the form of arch, the rules for ordering the stone, and the construction of the templates will be
found of considerable use. We commend the book to the engineering profession."—*Building News.*

"Will be regarded by civil engineers as of the utmost value, and calculated to save much time
and obviate many mistakes."—*Colliery Guardian.*

Girders, Strength of.

GRAPHIC TABLE FOR FACILITATING THE COMPUTA-
TION OF THE WEIGHTS OF WROUGHT IRON AND STEEL
GIRDERS, etc., for Parliamentary and other Estimates. By J. H. WATSON
BUCK, M.Inst.C.E. On a Sheet, 2s.6d.

River Engineering.

RIVER BARS: *The Causes of their Formation, and their Treatment by " Induced Tidal Scour;"* with a Description of the Successful Reduction by this Method of the Bar at Dublin. By I. J. MANN, Assist. Eng. to the Dublin Port and Docks Board. Royal 8vo, 7s. 6d. cloth.

" We recommend all interested in harbour works—and, indeed, those concerned in the improvements of rivers generally—to read Mr. Mann's interesting work on the treatment of river bars."—*Engineer.*

Trusses.

TRUSSES OF WOOD AND IRON. *Practical Applications of Science in Determining the Stresses, Breaking Weights, Safe Loads, Scantlings, and Details of Construction,* with Complete Working Drawings. By WILLIAM GRIFFITHS, Surveyor, Assistant Master. Tranmere School of Science and Art. Oblong 8vo, 4s. 6d. cloth.

" This handy little book enters so minutely into every detail connected with the construction of roof trusses, that no student need be ignorant of these matters."—*Practical Engineer.*

Railway Working.

SAFE RAILWAY WORKING. *A Treatise on Railway Accidents: Their Cause and Prevention; with a Description of Modern Appliances and Systems.* By CLEMENT E. STRETTON, C.E., Vice-President and Consulting Engineer, Amalgamated Society of Railway Servants. With Illustrations and Coloured Plates. Second Edition, Enlarged. Crown 8vo, 3s. 6d. cloth. [*Just published.*

" A book for the engineer, the directors, the managers ; and, in short, all who wish for information on railway matters will find a perfect encyclopædia in 'Safe Railway Working.'"—*Railway Review.*

" We commend the remarks on railway signalling to all railway managers, especially where a uniform code and practice is advocated."—*Herepath's Railway Journal.*

" The author may be congratulated on having collected, in a very convenient form, much valuable information on the principal questions affecting the safe working of railways."—*Railway Engineer.*

Field-Book for Engineers.

THE ENGINEER'S, MINING SURVEYOR'S, AND CONTRACTOR'S FIELD-BOOK. Consisting of a Series of Tables, with Rules, Explanations of Systems, and use of Theodolite for Traverse Surveying and Plotting the Work with minute accuracy by means of Straight Edge and Set Square only ; Levelling with the Theodolite, Casting-out and Reducing Levels to Datum, and Plotting Sections in the ordinary manner ; setting-out Curves with the Theodolite by Tangential Angles and Multiples, with Right and Left-hand Readings of the Instrument: Setting-out Curves without Theodolite, on the System of Tangential Angles by sets of Tangents and Off-sets ; and Earthwork Tables to 80 feet deep, calculated for every 6 inches in depth. By W. DAVIS HASKOLL, C.E. With numerous Woodcuts. Fourth Edition, Enlarged. Crown 8vo, 12s. cloth.

"The book is very handy : the separate tables of sines and tangents to every minute will make it useful for many other purposes, the genuine traverse tables existing all the same."—*Athenæum.*

" Every person engaged in engineering field operations will estimate the importance of such a work and the amount of valuable time which will be saved by reference to a set of reliable tables prepared with the accuracy and fulness of those given in this volume."—*Railway News.*

Earthwork, Measurement of.

A MANUAL ON EARTHWORK. By ALEX. J. S. GRAHAM, C.E. With numerous Diagrams. Second Edition. 18mo, 2s. 6d. cloth

"A great amount of practical information, very admirably arranged, and available for rough estimates, as well as for the more exact calculations required in the engineer's and contractor's offices."—*Artizan.*

Strains in Ironwork.

THE STRAINS ON STRUCTURES OF IRONWORK; with Practical Remarks on Iron Construction. By F. W. SHEILDS, M.Inst.C.E. Second Edition, with 5 Plates. Royal 8vo, 5s. cloth.

The student cannot find a better little book on this subject."—*Engineer.*

Cast Iron and other Metals, Strength of.

A PRACTICAL ESSAY ON THE STRENGTH OF CAST IRON AND OTHER METALS. By THOMAS TREDGOLD, C.E. Fifth Edition, including HODGKINSON's Experimental Researches. 8vo, 12s. cloth.

ARCHITECTURE, BUILDING, etc.

Construction.

THE SCIENCE OF BUILDING : *An Elementary Treatise on the Principles of Construction.* By E. WYNDHAM TARN, M.A., Architect. Third Edition, Enlarged, with 59 Engravings. Fcap. 8vo, 4s. cloth.

"A very valuable book, which we strongly recommend to all students."—*Builder.*
"No architectural student should be without this handbook."—*Architect.*

Villa Architecture.

A HANDY BOOK OF VILLA ARCHITECTURE : *Being a Series of Designs for Villa Residences in various Styles.* With Outline Specifications and Estimates. By C. WICKES, Author of "The Spires and Towers of England," &c. 61 Plates, 4to, £1 11s. 6d. half-morocco, gilt edges.

"The whole of the designs bear evidence of their being the work of an artistic architect, and they will prove very valuable and suggestive."—*Building News.*

Text-Book for Architects.

THE ARCHITECT'S GUIDE: *Being a Text-Book of Useful Information for Architects, Engineers, Surveyors, Contractors, Clerks of Works, &c. &c.* By FREDERICK ROGERS, Architect, Author of "Specifications for Practical Architecture," &c. Second Edition, Revised and Enlarged. With numerous Illustrations. Crown 8vo, 6s. cloth.

"As a text-book of useful information for architects, engineers, surveyors, &c., it would be hard to find a handier or more complete little volume."—*Standard.*
"A young architect could hardly have a better guide-book."—*Timber Trades Journal.*

Taylor and Cresy's Rome.

THE ARCHITECTURAL ANTIQUITIES OF ROME. By the late G. L. TAYLOR, Esq., F.R.I.B.A., and EDWARD CRESY, Esq. New Edition, thoroughly Revised by the Rev. ALEXANDER TAYLOR, M.A. (son of the late G. L. Taylor, Esq.), Fellow of Queen's College, Oxford, and Chaplain of Gray's Inn. Large folio, with 130 Plates, half-bound, £3 3s.

"Taylor and Cresy's work has from its first publication been ranked among those professional books which cannot be bettered. . . . It would be difficult to find examples of drawings, even among those of the most painstaking students of Gothic, more thoroughly worked out than are the one hundred and thirty plates in this volume."—*Architect.*

Linear Perspective.

ARCHITECTURAL PERSPECTIVE : The whole Course and Operations of the Draughtsman in Drawing a Large House in Linear Perspective. Illustrated by 39 Folding Plates. By F. O. FERGUSON. Demy 8vo, 3s. 6d. boards. [*Just published.*

Architectural Drawing.

PRACTICAL RULES ON DRAWING, *for the Operative Builder and Young Student in Architecture.* By GEORGE PYNE. With 14 Plates, 4to, 7s. 6d. boards.

Sir Wm. Chambers on Civil Architecture.

THE DECORATIVE PART OF CIVIL ARCHITECTURE. By Sir WILLIAM CHAMBERS, F.R.S. With Portrait, Illustrations, Notes, and an Examination of Grecian Architecture, by JOSEPH GWILT, F.S.A. Revised and Edited by W. H. LEEDS, with a Memoir of the Author. 66 Plates, 4to, 21s. cloth.

House Building and Repairing.

THE HOUSE-OWNER'S ESTIMATOR ; or, What will it Cost to Build, Alter, or Repair? A Price Book adapted to the Use of Unprofessional People, as well as for the Architectural Surveyor and Builder. By JAMES D. SIMON, A.R.I.B.A. Edited and Revised by FRANCIS T. W. MILLER, A.R.I.B.A. With numerous Illustrations. Fourth Edition, Revised. Crown 8vo, 3s. 6d. cloth.

"In two years it will repay its cost a hundred times over."—*Field.*

Cottages and Villas.

COUNTRY AND SUBURBAN COTTAGES AND VILLAS: How to Plan and Build Them. Containing 33 Plates, with Introduction, General Explanations, and Description of each Plate. By JAMES W. BOGUE, Architect, Author of "Domestic Architecture," &c. 4to, 10s. 6d. cloth.

The New Builder's Price Book, 1891.

LOCKWOOD'S BUILDER'S PRICE BOOK FOR 1891. A Comprehensive Handbook of the Latest Prices and Data for Builders, Architects, Engineers and Contractors. Re constructed, Re-written and Greatly Enlarged. By FRANCIS T. W. MILLER, 640 closely-printed pages, crown 8vo, 4s. cloth. [*Just published.*

" This book is a very useful one, and should find a place in every English office connected with the building and engineering professions."—*Industries.*

" This Price Book has been set up in new type. . . . Advantage has been taken of the transformation to add much additional information, and the volume is now an excellent book of reference."—*Architect.*

" In its new and revised form this Price Book is what a work of this kind should be—comprehensive, reliable, well arranged, legible and well b and.'—*British Architect.*

" A work of established reputation."—*Athenæum.*

" This very useful handbook is well written, exceedingly clear in its explanations and great care has evidently been taken to ensure accuracy."—*Morning Advertiser*

Designing, Measuring, and Valuing.

THE STUDENT'S GUIDE to the PRACTICE of MEASURING AND VALUING ARTIFICERS' WORKS. Containing Directions for taking Dimensions, Abstracting the same, and bringing the Quantities into Bill, with Tables of Constants for Valuation of Labour, and for the Calculation of Areas and Solidities. Originally edited by EDWARD DOBSON, Architect. With Additions on Mensuration and Construction, and a New Chapter on Dilapidations, Repairs, and Contracts, by E. WYNDHAM TARN, M.A. Sixth Edition, including a Complete Form of a Bill of Quantities. With 8 Plates and 63 Woodcuts. Crown 8vo, 7s. 6d. cloth.

" Well fulfils the promise of its title-page, and we can thoroughly recommend it to the class for whose use it has been compiled. Mr. Tarn's additions and revisions have much increased the usefulness of the work, and have especially augmented its value to students."—*Engineering.*

" This edition will be found the most complete treatise on the principles of measuring and valuing artificers' work that has yet been published."—*Building News.*

Pocket Estimator and Technical Guide.

THE POCKET TECHNICAL GUIDE, MEASURER AND ESTIMATOR FOR BUILDERS AND SURVEYORS. Containing Technical Directions for Measuring Work in all the Building Trades, Complete Specifications for Houses, Roads, and Drains, and an easy Method of Estimating the parts of a Building collectively. By A. C. BEATON, Author of "Quantities and Measurements," &c. Fifth Edition. With 53 Woodcuts, waistcoat-pocket size, 1s. 6d. gilt edges.

" No builder, architect, surveyor, or valuer should be without his ' Beaton."—*Building News.*

" Contains an extraordinary amount of information in daily requisition in measuring and estimating. Its presence in the pocket will save valuable time and trouble."—*Building World.*

Donaldson on Specifications.

THE HANDBOOK OF SPECIFICATIONS; or, Practical Guide to the Architect, Engineer, Surveyor, and Builder, in drawing up Specifications and Contracts for Works and Constructions. Illustrated by Precedents of Buildings actually executed by eminent Architects and Engineers. By Professor T. L. DONALDSON, P.R.I.B.A., &c. New Edition, in One large Vol., 8vo, with upwards of 1,000 pages of Text, and 33 Plates, £1 11s. 6d. cloth.

" In this work forty-four specifications of executed works are given, including the specifications for parts of the new Houses of Parliament, by Sir Charles Barry, and for the new Royal Exchange, by Mr. Tite, M.P. The latter, in particular, is a very complete and remarkable document. It embodies, to a great extent, as Mr. Donaldson mentions, 'the bill of quantities with the description of the works.' . . . It is valuable as a record, and more valuable still as a book of precedents. . . . Suffice it to say that Donaldson's 'Handbook of Specifications must be bought by all architects."—*Builder.*

Bartholomew and Rogers' Specifications.

SPECIFICATIONS FOR PRACTICAL ARCHITECTURE. A Guide to the Architect, Engineer, Surveyor, and Builder. With an Essay on the Structure and Science of Modern Buildings. Upon the Basis of the Work by ALFRED BARTHOLOMEW, thoroughly Revised, Corrected, and greatly added to by FREDERICK ROGERS, Architect. Second Edition, Revised, with Additions. With numerous Illustrations, medium 8vo, 15s. cloth.

" The collection of specifications prepared by Mr. Rogers on the basis of Bartholomew's work is too well known to need any recommendation from us. It is one of the books with which every young architect must be equipped : for time has shown that the specifications cannot be set aside through any defect in them."—*Architect.*

Building : Civil and Ecclesiastical.

A BOOK ON BUILDING, Civil and Ecclesiastical, including Church Restoration ; with the Theory of Domes and the Great Pyramid, &c. By Sir EDMUND BECKETT, Bart., LL.D., F.R.A.S., Author of "Clocks and Watches, and Bells," &c. Second Edition, Enlarged. Fcap. 8vo, 5s. cloth.

" A book which is always amusing and nearly always instructive. The style throughout is in the highest degree condensed and epigrammatic."—*Times.*

Ventilation of Buildings.

VENTILATION. A Text Book to the Practice of the Art of Ventilating Buildings. With a Chapter upon Air Testing. By W. P. BUCHAN, R.P., Sanitary and Ventilating Engineer, Author of "Plumbing," &c. With 170 Illustrations. 12mo, 4s. cloth boards. [*Just published.*

The Art of Plumbing.

PLUMBING. A Text Book to the Practice of the Art or Craft of the Plumber, with Supplementary Chapters on House Drainage, embodying the latest Improvements. By WILLIAM PATON BUCHAN, R.P., Sanitary Engineer and Practical Plumber. Fifth Edition, Enlarged to 370 pages, and 380 Illustrations. 12mo, 4s. cloth boards.

"A text book which may be safely put in the hands of every young plumber, and which will also be found useful by architects and medical professors."—*Builder.*

" A valuable text book, and the only treatise which can be regarded as a really reliable manual of the plumber's art."—*Building News.*

Geometry for the Architect, Engineer, etc.

PRACTICAL GEOMETRY, for the Architect, Engineer and Mechanic. Giving Rules for the Delineation and Application of various Geometrical Lines, Figures and Curves. By E. W. TARN, M.A., Architect, Author of "The Science of Building," &c. Second Edition. With 172 Illustrations, demy 8vo, 9s. cloth.

"No book with the same objects in view has ever been published in which the clearness of the rules laid down and the illustrative diagrams have been so satisfactory."—*Scotsman.*

The Science of Geometry.

THE GEOMETRY OF COMPASSES ; or, Problems Resolved by the mere Description of Circles, and the use of Coloured Diagrams and Symbols. By OLIVER BYRNE. Coloured Plates. Crown 8vo, 3s. 6d. cloth.

" The treatise is a good one, and remarkable—like all Mr. Byrne's contributions to the science of geometry—for the lucid character of its teaching."—*Building News.*

DECORATIVE ARTS, etc.

Woods and Marbles (Imitation of).

SCHOOL OF PAINTING FOR THE IMITATION OF WOODS AND MARBLES, as Taught and Practised by A. R. VAN DER BURG and P. VAN DER BURG, Directors of the Rotterdam Painting Institution. Royal folio, 18½ by 12¼ in., Illustrated with 24 full-size Coloured Plates; also 12 plain Plates, comprising 154 Figures. Second and Cheaper Edition. Price £1 11s. 6d.

List of Plates.

1. Various Tools required for Wood Painting—2, 3. Walnut : Preliminary Stages of Graining and Finished Specimen—4. Tools used for Marble Painting and Method of Manipulation—5, 6. St. Remi Marble : Earlier Operations and Finished Specimen—7. Methods of Sketching different Grains, Knots, &c.—8, 9. Ash : Preliminary Stages and Finished Specimen—10. Methods of Sketching Marble Grains—11, 12. Breche Marble : Preliminary Stages of Working and Finished Specimen—13. Maple : Methods of Producing the different Grains—14, 15. Bird's-eye Maple : Preliminary Stages and Finished Specimen—16. Methods of Sketching the different Species of White Marble—17, 18. White Marble : Preliminary Stages of Process and Finished Specimen—19. Mahogany : Specimens of various Grains and Methods of Manipulation—20, 21. Mahogany : Earlier Stages and Finished Specimen—22, 23, 24. Sienna Marble : Varieties of Grain, Preliminary Stages and Finished Specimen—25, 26, 27. Juniper Wood : Methods of producing Grain, &c.; Preliminary Stages and Finished Specimen—28, 29, 30. Vert de Mer Marble : Varieties of Grain and Methods of Working Unfinished and Finished Specimens—31, 32, 33. Oak : Varieties of Grain, Tools Employed, and Methods of Manipulation, Preliminary Stages and Finished Specimen—34, 35, 36. Waulsort Marble : Varieties of Grain, Unfinished and Finished Specimens.

** OPINIONS OF THE PRESS.

" Those who desire to attain skill in the art of painting woods and marbles will find advantage in consulting this book. . . . Some of the Working Men's Clubs should give their young men the opportunity to study it."—*Builder.*

"A comprehensive guide to the art. The explanations of the processes, the manipulation and management of the colours, and the beautifully executed plates will not be the least valuable to the student who aims at making his work a faithful transcript of nature."—*Building News.*

House Decoration.

ELEMENTARY DECORATION. A Guide to the Simpler Forms of Everyday Art, as applied to the Interior and Exterior Decoration of Dwelling Houses, &c. By JAMES W. FACEY, Jun. With 68 Cuts. 12mo, 2s. cloth limp.

PRACTICAL HOUSE DECORATION : A Guide to the Art of Ornamental Painting, the Arrangement of Colours in Apartments, and the principles of Decorative Design. With some Remarks upon the Nature and Properties of Pigments. By JAMES WILLIAM FACEY, Author of " Elementary Decoration," &c. With numerous Illustrations. 12mo, 2s. 6d. cloth limp.

N.B.—The above Two Works together in One Vol., strongly half-bound, 5s.

Colour.

A GRAMMAR OF COLOURING. Applied to Decorative Painting and the Arts. By GEORGE FIELD. New Edition, Revised, Enlarged, and adapted to the use of the Ornamental Painter and Designer. By ELLIS A. DAVIDSON. With New Coloured Diagrams and Engravings. 12mo, 3s. 6d. cloth boards.

"The book is a most useful *résumé* of the properties of pigments."—*Builder.*

House Painting, Graining, etc.

HOUSE PAINTING, GRAINING, MARBLING, AND SIGN WRITING, A Practical Manual of. By ELLIS A. DAVIDSON. Sixth Edition. With Coloured Plates and Wood Engravings. 12mo, 6s. cloth boards.

" A mass of information, of use to the amateur and of value to the practical man."—*Eng Mechanic.*
"Simply invaluable to the youngster entering upon this particular calling, and highly service-able to the man who is practising it."—*Furniture Gazette.*

Decorators, Receipts for.

THE DECORATOR'S ASSISTANT : A Modern Guide to Decorative Artists and Amateurs, Painters, Writers, Gilders, &c. Containing upwards of 600 Receipts, Rules and Instructions ; with a variety of Information for General Work connected with every Class of Interior and Exterior Decorations, &c. Fourth Edition, Revised. 152 pp., crown 8vo, 1s. in wrapper.

" Full of receipts of value to decorators, painters, gilders, &c. The book contains the gist of larger treatises on colour and technical processes. It would be difficult to meet with a work so full of varied information on the painter's art."—*Building News.*
" We recommend the work to all who, whether for pleasure or profit, require a guide to decoration."—*Plumber and Decorator.*

Moyr Smith on Interior Decoration.

ORNAMENTAL INTERIORS, ANCIENT AND MODERN. By J. MOYR SMITH. Super-royal 8vo, with 32 full-page Plates and numerous smaller Illustrations, handsomely bound in cloth, gilt top, price 18s.

" The book is well illustrated and handsomely got up, and contains some true criticism and a good many good examples of decorative treatment."—*The Builder.*
" This is the most elaborate and beautiful work on the artistic decoration of interiors that we have seen. . . . The scrolls, panels and other designs from the author's own pen are very beautiful and chaste ; but he takes care that the designs of other men shall figure even more than his own."—*Liverpool Albion.*
" To all who take an interest in elaborate domestic ornament this handsome volume will be welcome."—*Graphic.*

British and Foreign Marbles.

MARBLE DECORATION and the Terminology of British and Foreign Marbles. A Handbook for Students. By GEORGE H. BLAGROVE, Author of " Shoring and its Application," &c. With 28 Illustrations. Crown 8vo, 3s. 6d. cloth.

" This most useful and much wanted handbook should be in the hands of every architect an builder."—*Building World.*
" It is an excellent manual for students, and interesting to artistic readers generally."—*Saturday Review.*
" A carefully and usefully written treatise ; the work is essentially practical."—*Scotsman.*

Marble Working, etc.

MARBLE AND MARBLE WORKERS : A Handbook for Architects, Artists, Masons and Students. By ARTHUR LEE, Author of " A Visit to Carrara," " The Working of Marble," &c. Small crown 8vo, 2s. cloth.

" A really valuable addition to the technical literature of architects and masons."—*Building News.*

DELAMOTTE'S WORKS ON ILLUMINATION AND ALPHABETS.

A PRIMER OF THE ART OF ILLUMINATION, for the Use of Beginners : with a Rudimentary Treatise on tbe Art, Practical Directions for its exercise, and Examples taken from Illuminated MSS., printed in Gold and Colours. By F. DELAMOTTE. New and Cheaper Edition. Small 4to, 6s. ornamental boards.

"The examples of ancient MSS. recommended to the student, which, with much good sense, the author chooses from collections accessible to all, are selected with judgment and knowledge, as well as taste."—*Athenæum.*

ORNAMENTAL ALPHABETS, Ancient and Mediæval, from the Eighth Century, with Numerals ; including Gothic, Church-Text, large and small, German, Italian, Arabesque, Initials for Illumination, Monograms, Crosses, &c. &c., for the use of Architectural and Engineering Draughtsmen, Missal Painters, Masons, Decorative Painters, Lithographers, Engravers, Carvers, &c. &c. Collected and Engraved by F. DELAMOTTE, and printed in Colours. New and Cheaper Edition. Royal 8vo, oblong, 2s. 6d. ornamental boards.

"For those who insert enamelled sentences round gilded chalices, who blazon shop legends over shop-doors, who letter church walls with pithy sentences from the Decalogue, this book will be useful."—*Athenæum.*

EXAMPLES OF MODERN ALPHABETS, Plain and Ornamental ; including German, Old English, Saxon, Italic, Perspective, Greek, Hebrew, Court Hand, Engrossing, Tuscan, Riband, Gothic, Rustic, and Arabesque ; with several Original Designs, and an Analysis of the Roman and Old English Alphabets, large and small, and Numerals, for the use of Draughtsmen, Surveyors, Masons, Decorative Painters, Lithographers, Engravers, Carvers. &c. Collected and Engraved by F. DELAMOTTE, and printed in Colours. New and Cheaper Edition. Royal 8vo, oblong, 2s. 6d. ornamental boards.

"There is comprised in it every possible shape into which the letters of the alphabet and numerals can be formed, and the talent which has been expended in the conception of the various plain and ornamental letters is wonderful."—*Standard.*

MEDIÆVAL ALPHABETS AND INITIALS FOR ILLUMI-NATORS. By F. G. DELAMOTTE. Containing 21 Plates and Illuminated Title, printed in Gold and Colours. With an Introduction by J. WILLIS BROOKS. Fourth and Cheaper Edition. Small 4to, 4s. ornamental boards.

"A volume in which the letters of the alphabet come forth glorified in gilding and all the colours of the prism interwoven and intertwined and intermingled."—*Sun.*

THE EMBROIDERER'S BOOK OF DESIGN. Containing Initials, Emblems, Cyphers, Monograms, Ornamental Borders, Ecclesiastical Devices, Mediæval and Modern Alphabets, and National Emblems. Collected by F. DELAMOTTE, and printed in Colours. Oblong royal 8vo, 1s. 6d. ornamental wrapper.

"The book will be of great assistance to ladies and young children who are endowed with the art of plying the needle in this most ornamental and useful pretty work."—*East Anglian Times.*

Wood Carving.

INSTRUCTIONS IN WOOD-CARVING, for Amateurs ; with Hints on Design. By A LADY. With Ten Plates. New and Cheaper Edition. Crown 8vo, 2s. in emblematic wrapper.

"The handicraft of the wood-carver, so well as a book can impart it, may be learnt from ' A Lady's' publication."—*Athenæum.*
"The directions given are plain and easily understood."—*English Mechanic.*

Glass Painting.

GLASS STAINING AND THE ART OF PAINTING ON GLASS. From the German of Dr. GESSERT and EMANUEL OTTO FROMBERG. With an Appendix on THE ART OF ENAMELLING. 12mo, 2s. 6d. cloth limp.

Letter Painting.

THE ART OF LETTER PAINTING MADE EASY. By JAMES GREIG BADENOCH. With 12 full-page Engravings of Examples, 1s. 6d. cloth limp.

"The system is a simple one, but quite original, and well worth the careful attention of letter painters. It can be easily mastered and remembered."—*Building News.*

CARPENTRY, TIMBER, etc.

Tredgold's Carpentry, Revised & Enlarged by Tarn.

THE ELEMENTARY PRINCIPLES OF CARPENTRY.
A Treatise on the Pressure and Equilibrium of Timber Framing, the Resistance of Timber, and the Construction of Floors, Arches, Roofs, Uniting Iron and Stone with Timber, &c. To which is added an Essay on the Nature and Properties of Timber, &c., with Descriptions of the kinds of Wood used in Building; also numerous Tables of the Scantlings of Timber for different purposes, the Specific Gravities of Materials, &c. By THOMAS TREDGOLD, C.E. With an Appendix of Specimens of Various Roofs of Iron and Stone, Illustrated. Seventh Edition, thoroughly revised and considerably enlarged by E. WYNDHAM TARN, M.A., Author of "The Science of Building," &c. With 61 Plates, Portrait of the Author, and several Woodcuts. In one large vol., 4to, price £1 5s. cloth.

"Ought to be in every architect's and every builder's library."—*Builder.*
"A work whose monumental excellence must commend it wherever skilful carpentry is concerned. The author's principles are rather confirmed than impaired by time. The additional plates are of great intrinsic value."—*Building News.*

Woodworking Machinery.

WOODWORKING MACHINERY: Its Rise, Progress, and Construction. With Hints on the Management of Saw Mills and the Economical Conversion of Timber. Illustrated with Examples of Recent Designs by leading English, French, and American Engineers. By M. POWIS BALE, A.M.Inst.C.E., M.I.M.E. Large crown 8vo, 12s. 6d. cloth.

"Mr. Bale is evidently an expert on the subject and he has collected so much information that his book is all-sufficient for builders and others engaged in the conversion of timber."—*Architect.*
"The most comprehensive compendium of wood-working machinery we have seen. The author is a thorough master of his subject."—*Building News.*
"The appearance of this book at the present time will, we should think, give a considerable impetus to the onward march of the machinist engaged in the designing and manufacture of wood-working machines. It should be in the office of every wood-working factory."—*English Mechanic.*

Saw Mills.

SAW MILLS : Their Arrangement and Management, and the Economical Conversion of Timber. (A Companion Volume to "Woodworking Machinery.") By M. POWIS BALE. With numerous Illustrations. Crown 8vo, 10s. 6d. cloth.

"The administration of a large sawing establishment is discussed, and the subject examined from a financial standpoint. We could not desire a more complete or practical treatise."—*Builder.*
"We highly recommend Mr. Bale's work to the attention and perusal of all those who are engaged in the art of wood conversion, or who are about building or remodelling saw-mills on improved principles."—*Building News.*

Carpentering.

THE CARPENTER'S NEW GUIDE ; or, Book of Lines for Carpenters; comprising all the Elementary Principles essential for acquiring a knowledge of Carpentry. Founded on the late PETER NICHOLSON'S Standard Work. A New Edition, Revised by ARTHUR ASHPITEL, F.S.A. Together with Practical Rules on Drawing, by GEORGE PYNE. With 74 Plates, 4to, £1 1s. cloth.

Handrailing and Stairbuilding.

A PRACTICAL TREATISE ON HANDRAILING : Showing New and Simple Methods for Finding the Pitch of the Plank, Drawing the Moulds, Bevelling, Jointing-up, and Squaring the Wreath. By GEORGE COLLINGS. Second Edition, Revised and Enlarged, to which is added A TREATISE ON STAIRBUILDING. With Plates and Diagrams. 12mo, 2s. 6d. cloth limp.

"Will be found of practical utility in the execution of this difficult branch of joinery."—*Builder.*
"Almost every difficult phase of this somewhat intricate branch of joinery is elucidated by the aid of plates and explanatory letterpress."—*Furniture Gazette.*

Circular Work.

CIRCULAR WORK IN CARPENTRY AND JOINERY : A Practical Treatise on Circular Work of Single and Double Curvature. By GEORGE COLLINGS, Author of "A Practical Treatise on Handrailing." Illustrated with numerous Diagrams. Second Edition. 12mo, 2s. 6d. cloth limp.

"An excellent example of what a book of this kind should be. Cheap in price, clear in definition and practical in the examples selected."—*Builder.*

Timber Merchant's Companion.

THE TIMBER MERCHANT'S AND BUILDER'S COM-
PANION. Containing New and Copious Tables of the Reduced Weight and
Measurement of Deals and Battens, of all sizes, from One to a Thousand
Pieces, and the relative Price that each size bears per Lineal Foot to any
given Price per Petersburg Standard Hundred; the Price per Cube Foot of
Square Timber to any given Price per Load of 50 Feet; the proportionate
Value of Deals and Battens by the Standard, to Square Timber by the Load
of 50 Feet; the readiest mode of ascertaining the Price of Scantling per
Lineal Foot of any size, to any given Figure per Cube Foot, &c. &c.   By
WILLIAM DOWSING. Fourth Edition, Revised and Corrected.   Cr. 8vo, 3s. cl.
"We are glad to see a fourth edition of these admirable tables, which for correctness and
simplicity of arrangement leave nothing to be desired."—*Timber Trades Journal.*
"An exceedingly well-arranged, clear, and concise manual of tables for the use of all who buy
or sell timber."—*Journal of Forestry.*

Practical Timber Merchant.

THE PRACTICAL TIMBER MERCHANT. Being a Guide
for the use of Building Contractors, Surveyors, Builders, &c., comprising
useful Tables for all purposes connected with the Timber Trade, Marks of
Wood, Essay on the Strength of Timber, Remarks on the Growth of Timber,
&c. By W. RICHARDSON. Fcap. 8vo, 3s. 6d. cloth.
"This handy manual contains much valuable information for the use of timber merchants,
builders, foresters, and all others connected with the growth, sale, and manufacture of timber."—
Journal of Forestry.

Timber Freight Book.

THE TIMBER MERCHANT'S, SAW MILLER'S, AND
IMPORTER'S FREIGHT BOOK AND ASSISTANT. Comprising Rules,
Tables, and Memoranda relating to the Timber Trade. By WILLIAM
RICHARDSON, Timber Broker; together with a Chapter on "SPEEDS OF SAW
MILL MACHINERY," by M. POWIS BALE, M.I.M.E., &c. 12mo, 3s. 6d. cl. boards.
"A very useful manual of rules, tables, and memoranda relating to the timber trade. We re-
commend it as a compendium of calculation to all timber measurers and merchants, and as supply-
ing a real want in the trade."—*Building News.*

Packing-Case Makers, Tables for.

PACKING-CASE TABLES; showing the number of Super-
ficial Feet in Boxes or Packing-Cases, from six inches square and upwards.
By W. RICHARDSON, Timber Broker.   Third Edition.   Oblong 4to, 3s. 6d. cl.
"Invaluable labour-saving tables."—*Ironmonger.*
"Will save much labour and calculation."—*Grocer.*

Superficial Measurement.

THE TRADESMAN'S GUIDE TO SUPERFICIAL MEA-
SUREMENT. Tables calculated from 1 to 200 inches in length, by 1 to 108
inches in breadth. For the use of Architects, Surveyors, Engineers, Timber
Merchants, Builders, &c. By JAMES HAWKINGS. Third Edition. Fcap.,
3s. 6d. cloth.
"A useful collection of tables to facilitate rapid calculation of surfaces. The exact area of any
surface of which the limits have been ascertained can be instantly determined. The book will be
found of the greatest utility to all engaged in building operations."—*Scotsman.*
"These tables will be found of great assistance to all who require to make calculations in superficial measurement."—*English Mechanic.*

Forestry.

THE ELEMENTS OF FORESTRY. Designed to afford In-
formation concerning the Planting and Care of Forest Trees for Ornament or
Profit, with Suggestions upon the Creation and Care of Woodlands. By F. B.
HOUGH. Large crown 8vo, 10s. cloth.

Timber Importer's Guide.

THE TIMBER IMPORTER'S, TIMBER MERCHANT'S AND
BUILDER'S STANDARD GUIDE. By RICHARD E. GRANDY. Compris-
ing an Analysis of Deal Standards, Home and Foreign, with Comparative
Values and Tabular Arrangements for fixing Nett Landed Cost on Baltic
and North American Deals, including all intermediate Expenses, Freight,
Insurance, &c. &c. Together with copious Information for the Retailer and
Builder. Third Edition, Revised. 12mo, 2s. cloth limp.
"Everything it pretends to be: built up gradually, it leads one from a forest to a treenail, and
throws in, as a makeweight, a host of material concerning bricks, columns, cisterns, &c."—*English
Mechanic.*

MARINE ENGINEERING, NAVIGATION, etc.

Chain Cables.

CHAIN CABLES AND CHAINS. Comprising Sizes and Curves of Links, Studs, &c., Iron for Cables and Chains, Chain Cable and Chain Making, Forming and Welding Links, Strength of Cables and Chains, Certificates for Cables, Marking Cables, Prices of Chain Cables and Chains, Historical Notes, Acts of Parliament, Statutory Tests, Charges for Testing, List of Manufacturers of Cables, &c. &c. By THOMAS W. TRAILL, F.E.R.N., M. Inst. C.E., Engineer Surveyor in Chief, Board of Trade, Inspector of Chain Cable and Anchor Proving Establishments, and General Superintendent, Lloyd's Committee on Proving Establishments. With numerous Tables, Illustrations and Lithographic Drawings. Folio, £2 2s. cloth, bevelled boards.

"It contains a vast amount of valuable information. Nothing seems to be wanting to make it a complete and standard work of reference on the subject."—*Nautical Magazine.*

Marine Engineering.

MARINE ENGINES AND STEAM VESSELS (A Treatise on). By ROBERT MURRAY, C.E. Eighth Edition, thoroughly Revised, with considerable Additions by the Author and by GEORGE CARLISLE, C.E., Senior Surveyor to the Board of Trade at Liverpool. 12mo, 5s. cloth boards.

"Well adapted to give the young steamship engineer or marine engine and boiler maker a general introduction into his practical work."—*Mechanical World.*

"We feel sure that this thoroughly revised edition will continue to be as popular in the future as it has been in the past, as, for its size, it contains more useful information than any similar treatise."—*Industries.*

The information given is both sound and sensible, and well qualified to direct young seagoing hands on the straight road to the extra chief's certificate. Most useful to surveyors, inspectors, draughtsmen, and all young engineers who take an interest in their profession."—
Glasgow Herald.

"An indispensable manual for the student of marine engineering."—*Liverpool Mercury.*

Pocket-Book for Naval Architects and Shipbuilders.

THE NAVAL ARCHITECT'S AND SHIPBUILDER'S POCKET-BOOK of Formulæ, Rules, and Tables, and MARINE ENGINEER'S AND SURVEYOR'S Handy Book of Reference. By CLEMENT MACKROW, Member of the Institution of Naval Architects, Naval Draughtsman. Fourth Edition, Revised. With numerous Diagrams, &c. Fcap., 12s. 6d. strongly bound in leather.

"Will be found to contain the most useful tables and formulæ required by shipbuilders, carefully collected from the best authorities, and put together in a popular and simple form."—*Engineer.*

"The professional shipbuilder has now, in a convenient and accessible form, reliable data for solving many of the numerous problems that present themselves in the course of his work."—*Iron.*

"There is scarcely a subject on which a naval architect or shipbuilder can require to refresh his memory which will not be found within the covers of Mr. Mackrow's book."—*English Mechanic.*

Pocket-Book for Marine Engineers.

A POCKET-BOOK OF USEFUL TABLES AND FOR-MULÆ FOR MARINE ENGINEERS. By FRANK PROCTOR, A.I.N.A. Third Edition. Royal 32mo, leather, gilt edges, with strap, 4s.

"We recommend it to our readers as going far to supply a long-felt want."—*Naval Science.*

"A most useful companion to all marine engineers."—*United Service Gazette.*

Introduction to Marine Engineering.

ELEMENTARY ENGINEERING: A Manual for Young Marine Engineers and Apprentices. In the Form of Questions and Answers on Metals, Alloys, Strength of Materials, Construction and Management of Marine Engines and Boilers, Geometry, &c. &c. With an Appendix of Useful Tables. By JOHN SHERREN BREWER, Government Marine Surveyor, Hong-kong. Small crown 8vo, 2s. cloth.

"Contains much valuable information for the class for whom it is intended, especially in the chapters on the management of boilers and engines."—*Nautical Magazine.*

"A useful introduction to the more elaborate text books."—*Scotsman.*

"To a student who has the requisite desire and resolve to attain a thorough knowledge, Mr. Brewer offers decidedly useful help."—*Athenæum.*

Navigation.

PRACTICAL NAVIGATION. Consisting of THE SAILOR'S SEA-BOOK, by JAMES GREENWOOD and W. H. ROSSER; together with the requisite Mathematical and Nautical Tables for the Working of the Problems, by HENRY LAW, C.E., and Professor J. R. YOUNG. Illustrated. 12mo, 7s. strongly half-bound.

MINING AND METALLURGY.

Metalliferous Mining in the United Kingdom.

BRITISH MINING : *A Treatise on the History, Discovery, Practical Development, and Future Prospects of Metalliferous Mines in the United Kingdom.* By ROBERT HUNT, F.R.S., Keeper of Mining Records; Editor of "Ure's Dictionary of Arts, Manufactures, and Mines," &c. Upwards of 950 pp., with 230 Illustrations. Second Edition, Revised. Super-royal 8vo, £2 2s. cloth.

"One of the most valuable works of reference of modern times. Mr. Hunt, as keeper of mining records of the United Kingdom, has had opportunities for such a task not enjoyed by anyone else, and has evidently made the most of them. . . . The language and style adopted are good, and the treatment of the various subjects laborious, conscientious, and scientific."—*Engineering.*

"The book is, in fact, a treasure-house of statistical information on mining subjects, and we know of no other work embodying so great a mass of matter of this kind. Were this the only merit of Mr. Hunt s volume, it would be sufficient to render it indispensable in this library of everyone interested in the development of the mining and metallurgical industries of this country."—*Athenæum.*

"A mass of information not elsewhere available, and of the greatest value to those who may be interested in our great mineral industries."—*Engineer.*

"A sound, business-like collection of interesting facts. . . . The amount of information Mr. Hunt has brought together is enormous. . . . The volume appears likely to convey more Instruction upon the subject than any work hitherto published."—*Mining Journal.*

Colliery Management.

THE COLLIERY MANAGER'S HANDBOOK : A Comprehensive Treatise on the Laying-out and Working of Collieries, Designed as a Book of Reference for Colliery Managers, and for the Use of Coal-Mining Students preparing for First-class Certificates. By CALEB PAMELY, Mining Engineer and Surveyor; Member of the North of England Institute of Mining and Mechanical Engineers; and Member of the South Wales Institute of Mining Engineers. With nearly 500 Plans, Diagrams, and other Illustrations. Medium 8vo, about 600 pages. Price £1 5s. strongly bound.
[*Just published.*

Coal and Iron.

THE COAL AND IRON INDUSTRIES OF THE UNITED KINGDOM. Comprising a Description of the Coal Fields, and of the Principal Seams of Coal, with Returns of their Produce and its Distribution, and Analyses of Special Varieties. Also an Account of the occurrence of Iron Ores in Veins or Seams; Analyses of each Variety; and a History of the Rise and Progress of Pig Iron Manufacture. By RICHARD MEADE, Assistant Keeper of Mining Records. With Maps. 8vo, £1 8s. cloth.

"The book is one which must find a place on the shelves of all interested in coal and iron production, and in the iron, steel, and other metallurgical industries."—*Engineer.*

"Of this book we may unreservedly say that it is the best of its class which we have ever met. . . . A book of reference which no one engaged in the iron or coal trades should omit from his library."—*Iron and Coal Trades Review.*

Prospecting for Gold and other Metals.

THE PROSPECTOR'S HANDBOOK : A Guide for the Prospector and Traveller in Search of Metal-Bearing or other Valuable Minerals. By J. W. ANDERSON, M.A. (Camb.), F.R.G.S., Author of "Fiji and New Caledonia." Fifth Edition, thoroughly Revised and Enlarged. Small crown 8vo, 3s. 6d. cloth.

"Will supply a much felt want, especially among Colonists, in whose way are so often thrown many mineralogical specimens the value of which it is difficult to determine."—*Engineer.*

"How to find commercial minerals, and how to identify them when they are found, are the leading points to which attention is directed. The author has managed to pack as much practical detail into his pages as would supply material for a book three times its size."—*Mining Journal.*

Mining Notes and Formulæ.

NOTES AND FORMULÆ FOR MINING STUDENTS. By JOHN HERMAN MERIVALE, M.A., Certificated Colliery Manager, Professor of Mining in the Durham College of Science, Newcastle-upon-Tyne. Third Edition, Revised and Enlarged. Small crown 8vo, 2s. 6d. cloth.

"Invaluable to anyone who is working up for an examination on mining subjects."—*Coal and Iron Trades Review.*

"The author has done his work in an exceedingly creditable manner, and has produced a book that will be of service to students, and those who are practically engaged in mining operations."—*Engineer.*

"A vast amount of technical matter of the utmost value to mining engineers, and of considerable interest to students."—*Schoolmaster.*

Explosives.

A HANDBOOK ON MODERN EXPLOSIVES. Being a Practical Treatise on the Manufacture and Application of Dynamite, Gun-Cotton, Nitro-Glycerine and other Explosive Compounds. Including the Manufacture of Collodion-Cotton. By M. EISSLER, Mining Engineer and Metallurgical Chemist, Author of "The Metallurgy of Gold," &c. With about 100 Illustrations. Crown 8vo, 10s. 6d. cloth.

"Useful not only to the miner, but also to officers of both services to whom blasting and the use of explosives generally may at any time become a necessary auxiliary." *Nature.*
"A veritable mine of information on the subject of explosives employed for military, mining and blasting purposes."—*Army and Navy Gazette.*
"The book is clearly written. Taken as a whole, we consider it an excellent little book and one that should be found of great service to miners and others who are engaged in work requiring the use of explosives."—*Athenæum.*

Gold, Metallurgy of.

THE METALLURGY OF GOLD: A Practical Treatise on the Metallurgical Treatment of Gold-bearing Ores. Including the Processes of Concentration and Chlorination, and the Assaying, Melting and Refining of Gold. By M. EISSLER, Mining Engineer and Metallurgical Chemist, formerly Assistant Assayer of the U. S. Mint, San Francisco. Third Edition, Revised and greatly Enlarged. With 187 Illustrations. Crown 8vo, 12s. 6d. cloth.

"This book thoroughly deserves its title of a 'Practical Treatise.' The whole process of gold milling, from the breaking of the quartz to the assay of the bullion is described in clear and orderly narrative and with much, but not too much, fulness of detail."—*Saturday Review.*
"The work is a storehouse of information and valuable data, and we strongly recommend it to all professional men engaged in the gold-mining industry."—*Mining Journal*

Silver, Metallurgy of.

THE METALLURGY OF SILVER: A Practical Treatise on the Amalgamation, Roasting and Lixiviation of Silver Ores, Including the Assaying, Melting and Refining of Silver Bullion. By M. EISSLER, Author of "The Metallurgy of Gold" Second Edition, Enlarged. With 150 Illustrations. Crown 8vo, 10s. 6d. cloth.   [*Just published.*

"A practical treatise, and a technical work which we are convinced will supply a long-felt want amongst practical men, and at the same time be of value to students and others indirectly connected with the industries."—*Mining Journal.*
"From first to last the book is thoroughly sound and reliable."—*Colliery Guardian.*
"For chemists, practical miners, assayers and investors alike, we do not know of any work on the subject so handy and yet so comprehensive."—*Glasgow Herald.*

Silver-Lead, Metallurgy of.

THE METALLURGY OF ARGENTIFEROUS LEAD: A Practical Treatise on the Smelting of Silver-Lead Ores and the Refining of Lead Bullion. Including Reports on various Smelting Establishments and Descriptions of Modern Furnaces and Plants in Europe and America. By M. EISSLER, M.E., Author of "The Metallurgy of Gold," &c. Crown 8vo. 400 pp., with numerous Illustrations, 12s. 6d. cloth.   [*Just published.*

Metalliferous Minerals and Mining.

TREATISE ON METALLIFEROUS MINERALS AND MINING. By D. C. DAVIES, F.G.S., Mining Engineer, &c., Author of "A Treatise on Slate and Slate Quarrying." Illustrated with numerous Wood Engravings. Fourth Edition, carefully Revised. Crown 8vo, 12s. 6d. cloth.

"Neither the practical miner nor the general reader interested in mines can have a better boo for his companion and his guide."—*Mining Journal.*   [*Mining World.*
"We are doing our readers a service in calling their attention to this valuable work."—
"As a history of the present state of mining throughout the world this book has a real value, and it supplies an actual want."—*Athenæum.*

Earthy Minerals and Mining.

A TREATISE ON EARTHY & OTHER MINERALS AND MINING. By D. C. DAVIES, F.G.S. Uniform with, and forming a Companion Volume to, the same Author's "Metalliferous Minerals and Mining." With 76 Wood Engravings. Second Edition. Crown 8vo, 12s. 6d. cloth.

"We do not remember to have met with any English work on mining matters that contains the same amount of information packed in equally convenient form."—*Academy.*
"We should be inclined to rank it as among the very best of the handy technical and trades manuals which have recently appeared."—*British Quarterly Review.*

Mineral Surveying and Valuing.

THE MINERAL SURVEYOR AND VALUER'S COMPLETE
GUIDE, *comprising a Treatise on Improved Mining Surveying and the Valuation of Mining Properties, with New Traverse Tables.* By WM. LINTERN,
Mining and Civil Engineer. Third Edition, with an Appendix on "Magnetic
and Angular Surveying," with Records of the Peculiarities of Needle Disturbances. With Four Plates of Diagrams, Plans, &c. 12mo, 4s. cloth.

" Mr. Lintern's book forms a valuable and thoroughly trustworthy guide."—*Iron and Coal Trades Review.*

" This new edition must be of the highest value to colliery surveyors, proprietors and managers."—*Colliery Guardian.*

Asbestos and its Uses.

ASBESTOS : *Its Properties, Occurrence and Uses.* With some
Account of the Mines of Italy and Canada. By ROBERT H. JONES. With
Eight Collotype Plates and other Illustrations. Crown 8vo, 12s. 6d. cloth.

" An interesting and invaluable work."—*Colliery Guardian.*

" We counsel our readers to get this exceedingly interesting work for themselves ; they will find in it much that is suggestive, and a great deal that is of immediate and practical usefulness."—*Builder.*

" A valuable addition to the architect's and engineer's library."—*Building News.*

Underground Pumping Machinery.

MINE DRAINAGE. Being a Complete and Practical Treatise
on Direct-Acting Underground Steam Pumping Machinery, with a Description of a large number of the best known Engines, their General Utility and
the Special Sphere of their Action, the Mode of their Application, and
their merits compared with other forms of Pumping Machinery. By STEPHEN
MICHELL. 8vo, 15s. cloth.

" Will be highly esteemed by colliery owners and lessees, mining engineers, and students generally who require to be acquainted with the best means of securing the drainage of mines. It is a most valuable work, and stands almost alone in the literature of steam pumping machinery."—*Colliery Guardian.*

" Much valuable information is given, so that the book is thoroughly worthy of an extensive circulation amongst practical men and purchasers of machinery."—*Mining Journal.*

Mining Tools.

A MANUAL OF MINING TOOLS. For the Use of Mine
Managers, Agents, Students, &c. By WILLIAM MORGANS, Lecturer on Practical Mining at the Bristol School of Mines. 12mo, 2s. 6d. cloth limp.

ATLAS OF ENGRAVINGS to Illustrate the above, containing 235 Illustrations of Mining Tools, drawn to scale. 4to, 4s. 6d. cloth.

" Students in the science of mining, and overmen, captains, managers, and viewers may gain practical knowledge and useful hints by the study of Mr. Morgans' manual."—*Colliery Guardian.*

" A valuable work, which will tend materially to improve our mining literature."—*Mining Journal.*

Coal Mining.

COAL AND COAL MINING : *A Rudimentary Treatise on.* By
the late Sir WARINGTON W. SMYTH, M.A., F.R.S., &c., Chief Inspector of the
Mines of the Crown. Seventh Edition, Revised and Enlarged. With
numerous Illustrations. 12mo, 4s. cloth boards.

" As an outline is given of every known coal-field in this and other countries, as well as of the principal methods of working, the book will doubtless interest a very large number of readers."—*Mining Journal.*

Subterraneous Surveying.

SUBTERRANEOUS SURVEYING, *Elementary and Practical
Treatise on,* with and without the Magnetic Needle. By THOMAS FENWICK,
Surveyor of Mines, and THOMAS BAKER, C.E. Illust. 12mo, 3s. cloth boards.

Granite Quarrying.

GRANITES AND OUR GRANITE INDUSTRIES. By
GEORGE F. HARRIS, F.G.S., Membre de la Société Belge de Géologie, Lecturer on Economic Geology at the Birkbeck Institution, &c. With Illustrations. Crown 8vo, 2s. 6d. cloth.

" A clearly and well-written manual for persons engaged or interested in the granite industry."
—*Scotsman.*

" An interesting work, which will be deservedly esteemed."—*Colliery Guardian.*

" An exceedingly interesting and valuable monograph on a subject which has hitherto received unaccountably little attention in the shape of systematic literary treatment."—*Scottish Leader.*

ELECTRICITY ELECTRICAL ENGINEERING, etc.

Electrical Engineering.

THE ELECTRICAL ENGINEER'S POCKET-BOOK OF
MODERN RULES, FORMULÆ, TABLES AND DATA. By H. R.
KEMPE, M.Inst.E.E., A.M.Inst C.E., Technical Officer Postal Telegraphs,
Author of "A Handbook of Electrical Testing," &c. With numerous Illus-
trations, royal 32mo, oblong, 5s. leather. *[Just published.*

" There is very little in the shape of formulæ or data which the electrician is likely to want
in a hurry which cannot be found in its pages."—*Practical Engineer.*

"A very useful book of reference for daily use in practical electrical engineering and its
various applications to the industries of the present day."—*Iron.*

" It is the best book of its kind."—*Electrical Engineer.*

"The Electrical Engineer's Pocket-Book is a good one."—*Flectrician.*

"Strongly recommended to those engaged in the various electrical industries."—*Electrical
Review.*

Electric Lighting.

ELECTRIC LIGHT FITTING : A Handbook for Working
Electrical Engineers, embodying Practical Notes on Installation Manage-
ment. By JOHN W. URQUHART, Electrician, Author of "Electric Light," &c.
With numerous Illustrations, crown 8vo, 5s. cloth. *[Just published.*

"This volume deals with what may be termed the mechanics of electric lighting, and is
addressed to men who are already engaged in the work or are training for it. The work traverses
a great deal of ground, and may be read as a sequel to the same author's useful work on ' Electric
Light.' '—*Electrician.*

" This is an attempt to state in the simplest language the precautions which should be adopted
in instaling the electric light, and to give information,for the guidance of those who have to run
the plant when installed. The book is well worth the perusal of the workmen for whom it is
written."—*Electrical Review.*

" Eminently practical and useful. . . . Ought to be in the hands of everyone in charge of
an electric light plant."—*Electrical Engineer.*

"A really capital book, which we have no hesitation in recommending to the notice of working
electricians and electrical engineers."—*Mechanical World.*

Electric Light.

ELECTRIC LIGHT : *Its Production and Use.* Embodying Plain
Directions for the Treatment of Dynamo-Electric Machines, Batteries,
Accumulators, and Electric Lamps. By J. W. URQUHART, C.E., Author of
" Electric Light Fitting," &c. Fourth Edition, Revised, with Large Additions
and 145 Illustrations. Crown 8vo, 7s. 6d. cloth. *[Just published.*

"The book is by far the best that we have yet met with on the subject."—*Athenæum.*

"It is the only work at present available which gives, in language intelligible for the most part
to the ordinary reader, a general but concise history of the means which have been adopted up to
the present time in producing the electric light."—*Metropolitan.*

"The book contains a general account of the means adopted in producing the electric light,
not only as obtained from voltaic or galvanic batteries, but treats at length of the dynamo-electric
machine in several of its forms."—*Colliery Guardian.*

Construction of Dynamos.

DYNAMO CONSTRUCTION : *A Practical Handbook for the Use
of Engineer Constructors and Electricians in Charge.* With Examples of
leading English, American and Continental Dynamos and Motors. By J. W.
URQUHART, Author of "Electric Light," &c. Crown 8vo, 7s. cloth.
[Just published.

'The author has produced a book for which a demand has long existed. The subject is
treated in a thoroughly practical manner. '—*Mechanical World.*

Dynamic Electricity and Magnetism.

THE ELEMENTS OF DYNAMIC ELECTRICITY AND
MAGNETISM By PHILIP ATKINSON, A.M., Ph.D. Crown 8vo. 400 pp.
With 120 Illustrations. 10s. 6d. cloth. *[just published.*

Text Book of Electricity.

THE STUDENT'S TEXT-BOOK OF ELECTRICITY. By
HENRY M. NOAD, Ph.D., F.R.S., F.C.S. New Edition, carefully Revised.
With an Introduction and Additional Chapters, by W. H. PREECE, M.I.C.E.,
Vice-President of the Society of Telegraph Engineers, &c. With 470 Illustra-
tions. Crown 8vo, 12s. 6d. cloth.

'We can recommend Dr. Noad's book for clear style, great range of subject, a good index,
d a plethora of woodcuts. Such collections as the present are indispensable."—*Athenæum.*

"An admirable text book for every student — beginner or advanced — of electricity."—
Engineering.

Electric Lighting.

THE ELEMENTARY PRINCIPLES OF ELECTRIC LIGHT-
ING. By ALAN A. CAMPBELL SWINTON, Associate I.E.E. Second Edition,
Enlarged and Revised. With 16 Illustrations. Crown 8vo, 1s. 6d. cloth.
"Anyone who desires a short and thoroughly clear exposition of the elementary principles of
electric-lighting cannot do better than read this little work."—*Bradford Observer.*

Electricity.

A MANUAL OF ELECTRICITY : Including Galvanism, Mag-
netism, Dia-Magnetism, Electro-Dynamics, Magno-Electricity, and the Electric
Telegraph. By HENRY M. NOAD, Ph.D., F.R.S., F.C.S. Fourth Edition.
With 500 Woodcuts. 8vo, £1 4s. cloth.
"It is worthy of a place in the library of every public institution."—*Mining Journal.*

Dynamo Construction.

HOW TO MAKE A DYNAMO : A Practical Treatise for Amateurs.
Containing numerous Illustrations and Detailed Instructions for Constructing a Small Dynamo, to Produce the Electric Light. By ALFRED CROFTS.
Third Edition, Revised and Enlarged. Crown 8vo, 2s. cloth.
"The instructions given in this unpretentious little book are sufficiently clear and explicit to
enable any amateur mechanic possessed of average skill and the usual tools to be found in an
amateur's workshop, to build a practical dynamo machine."—*Electrician.*

NATURAL SCIENCE, etc.

Pneumatics and Acoustics.

PNEUMATICS : including Acoustics and the Phenomena of Wind
Currents, for the Use of Beginners. By CHARLES TOMLINSON, F.R.S.
F.C.S., &c. Fourth Edition, Enlarged. 12mo, 1s. 6d. cloth.
"Beginners in the study of this important application of science could not have a better manual "
—*Scotsman.* " A valuable and suitable text-book for students of Acoustics and the Pheno-
mena of Wind Currents."—*Schoolmaster.*

Conchology.

A MANUAL OF THE MOLLUSCA : Being a Treatise on Recent
and Fossil Shells. By S. P. WOODWARD, A.L.S., F.G.S., late Assistant
Palæontologist in the British Museum. With an Appendix on Recent and
Fossil Conchological Discoveries, by RALPH TATE, A.L.S., F.G.S. Illustrated
by A. N. WATERHOUSE and JOSEPH WILSON LOWRY. With 23 Plates and
upwards of 300 Woodcuts. Reprint of Fourth Ed., 1880. Cr. 8vo, 7s. 6d. cl.
"A most valuable storehouse of conchological and geological information."—*Science Gossip.*

Geology.

RUDIMENTARY TREATISE ON GEOLOGY, PHYSICAL
AND HISTORICAL. Consisting of "Physical Geology," which sets forth
the leading Principles of the Science; and "Historical Geology," which
treats of the Mineral and Organic Conditions of the Earth at each successive
epoch, especial reference being made to the British Series of Rocks. By
RALPH TATE, A.L.S., F.G.S., &c. With 250 Illustrations. 12mo, 5s. cloth.
"The fulness of the matter has elevated the book into a manual. Its information is exhaustive
and well arranged."—*School Board Chronicle.*

Geology and Genesis.

THE TWIN RECORDS OF CREATION ; or, Geology and
Genesis: their Perfect Harmony and Wonderful Concord. By GEORGE W.
VICTOR LE VAUX. Numerous Illustrations. Fcap. 8vo, 5s. cloth.
"A valuable contribution to the evidences of Revelation, and disposes very conclusively of the
arguments of those who would set God's Works against God's Word."—*The Rock.*

The Constellations.

STAR GROUPS: A Student's Guide to the Constellations. By
J. ELLARD GORE, F.R.A.S., M.R.I.A., &c., Author of "The Scenery of the
Heavens." With 30 Maps. Small 4to, 5s. cloth, silvered. [*Just published.*

Astronomy.

ASTRONOMY. By the late Rev. ROBERT MAIN, M.A., F.R.S.,
formerly Radcliffe Observer at Oxford. Third Edition, Revised and Cor-
rected to the present time, by W. T. LYNN, B.A., F.R.A.S. 12mo, 2s. cloth.
"A sound and simple treatise, very carefully edited, and a capital book for beginners."—
Knowledge. [*tional Times.*
" Accurately brought down to the requirements of the present time by Mr. Lynn."—*Educa-*

DR. LARDNER'S COURSE OF NATURAL PHILOSOPHY.

THE HANDBOOK OF MECHANICS. Enlarged and almost re-written by Benjamin Loewy, F.R.A.S. With 378 Illustrations. Post 8vo, 6s. cloth.

"The perspicuity of the original has been retained, and chapters which had become obsolete have been replaced by others of more modern character. The explanations throughout are studiously popular, and care has been taken to show the application of the various branches of physics to the industrial arts, and to the practical business of life."—*Mining Journal.*

"Mr. Loewy has carefully revised the book, and brought it up to modern requirements."—*Nature.*

"Natural philosophy has had few exponents more able or better skilled in the art of popularising the subject than Dr. Lardner ; and Mr. Loewy is doing good service in fitting this treatise, and the others of the series, for use at the present time."—*Scotsman.*

THE HANDBOOK OF HYDROSTATICS AND PNEUMATICS. New Edition, Revised and Enlarged, by Benjamin Loewy, F.R.A.S. With 236 Illustrations. Post 8vo, 5s. cloth.

"For those 'who desire to attain an accurate knowledge of physical science without the profound methods of mathematical investigation,' this work is not merely intended, but well adapted."—*Chemical News.*

"The volume before us has been carefully edited, augmented to nearly twice the bulk of the former edition, and all the most recent matter has been added. . . . It is a valuable text-book."—*Nature.*

"Candidates for pass examinations will find it, we think, specially suited to their requirements."
English Mechanic.

THE HANDBOOK OF HEAT. Edited and almost entirely re-written by Benjamin Loewy, F.R.A.S., &c. 117 Illustrations. Post 8vo, 6s. cloth.

"The style is always clear and precise, and conveys instruction without leaving any cloudiness or lurking doubts behind."—*Engineering.*

"A most exhaustive book on the subject on which it treats, and is so arranged that it can be understood by all who desire to attain an accurate knowledge of physical science. . . . . Mr. Loewy has included all the latest discoveries in the varied laws and effects of heat."—*Standard.*

"A complete and handy text-book for the use of students and general readers."—*English Mechanic.*

THE HANDBOOK OF OPTICS. By Dionysius Lardner, D.C.L., formerly Professor of Natural Philosophy and Astronomy in University College, London. New Edition. Edited by T. Olver Harding, B.A. Lond., of University College, London. With 298 Illustrations. Small 8vo, 448 pages, 5s. cloth.

"Written by one of the ablest English scientific writers, beautifully and elaborately illustrated."
Mechanic's Magazine.

THE HANDBOOK OF ELECTRICITY, MAGNETISM, AND ACOUSTICS. By Dr. Lardner. Ninth Thousand. Edit. by George Carey Foster, B.A., F.C.S. With 400 Illustrations. Small 8vo, 5s. cloth.

"The book could not have been entrusted to anyone better calculated to preserve the terse and lucid style of Lardner, while correcting his errors and bringing up his work to the present state of scientific knowledge."—*Popular Science Review.*

—— —— ——

THE HANDBOOK OF ASTRONOMY. Forming a Companion to the " Handbook of Natural Philosophy." By Dionysius Lardner, D.C.L., formerly Professor of Natural Philosophy and Astronomy in University College, London. Fourth Edition. Revised and Edited by Edwin Dunkin, F.R.A.S., Royal Observatory, Greenwich. With 38 Plates and upwards of 100 Woodcuts. In One Vol., small 8vo, 550 pages, 9s. 6d. cloth.

"Probably no other book contains the same amount of information in so compendious and well-arranged a form—certainly none at the price at which this is offered to the public."—*Athenæum.*

"We can do no other than pronounce this work a most valuable manual of astronomy, and we strongly recommend it to all who wish to acquire a general—but at the same time correct—acquaintance with this sublime science."—*Quarterly Journal of Science.*

"One of the most deservedly popular books on the subject . . . We would recommend not only the student of the elementary principles of the science, but he who aims at mastering the higher and mathematical branches of astronomy, not to be without this work beside him."—*Practical Magazine.*

Dr. Lardner's Electric Telegraph.

THE ELECTRIC TELEGRAPH. By Dr. Lardner. Revised and Re-written by E. B. Bright, F.R.A.S. 140 Illustrations. Small 8vo, 2s. 6d. cloth.

"One of the most readable books extant on the Electric Telegraph."—*English Mechanic.*

DR. LARDNER'S MUSEUM OF SCIENCE AND ART.

THE MUSEUM OF SCIENCE AND ART. Edited by
DIONYSIUS LARDNER, D.C.L., formerly Professor of Natural Philosophy and
Astronomy in University College, London. With upwards of 1,200 Engravings on Wood. In 6 Double Volumes, £1 1s., in a new and elegant cloth bind-
ing; or handsomely bound in half-morocco, 31s. 6d.

*** OPINIONS OF THE PRESS.

"This series, besides affording popular but sound instruction on scientific subjects, with which
the humblest man in the country ought to be acquainted, also undertakes that teaching of 'Common Things' which every well-wisher of his kind is anxious to promote. Many thousand copies of
this serviceable publication have been printed, in the belief and hope that the desire for instruction
and improvement widely prevails; and we have no fear that such enlightened faith will meet with
disappointment."—*Times*.

"A cheap and interesting publication, alike informing and attractive. The papers combine
subjects of importance and great scientific knowledge, considerable inductive powers, and a
popular style of treatment."—*Spectator*.

"The 'Museum of Science and Art' is the most valuable contribution that has ever been
made to the Scientific Instruction of every class of society."—Sir DAVID BREWSTER, in the
North British Review.

"Whether we consider the liberality and beauty of the illustrations, the charm of the writing,
or the durable interest of the matter, we must express our belief that there is hardly to be found
among the new books one that would be welcomed by people of so many ages and classes as a
valuable present."—*Examiner*.

*** *Separate books formed from the above, suitable for Workmen's Libraries,*
Science Classes, etc.

Common Things Explained. Containing Air, Earth, Fire, Water, Time,
Man, the Eye, Locomotion, Colour, Clocks and Watches, &c. 233 Illustrations, cloth gilt, 5s.

The Microscope. Containing Optical Images, Magnifying Glasses, Origin
and Description of the Microscope, Microscopic Objects, the Solar Microscope, Microscopic Drawing and Engraving, &c. 147 Illustrations, cloth
gilt, 2s.

Popular Geology. Containing Earthquakes and Volcanoes, the Crust of
the Earth, &c. 201 Illustrations, cloth gilt, 2s. 6d.

Popular Physics. Containing Magnitude and Minuteness, the Atmosphere, Meteoric Stones, Popular Fallacies, Weather Prognostics, the
Thermometer, the Barometer, Sound, &c. 85 Illustrations, cloth gilt, 2s. 6d.

Steam and its Uses. Including the Steam Engine, the Locomotive, and
Steam Navigation. 89 Illustrations, cloth gilt, 2s.

Popular Astronomy. Containing How to observe the Heavens—The
Earth, Sun, Moon, Planets, Light, Comets, Eclipses, Astronomical Influ-
ences, &c. 182 Illustrations, 4s. 6d.

The Bee and White Ants : Their Manners and Habits. With Illustrations of Animal Instinct and Intelligence. 135 Illustrations, cloth gilt, 2s.

The Electric Telegraph Popularized. To render intelligible to all who
can Read, irrespective of any previous Scientific Acquirements, the various
forms of Telegraphy in Actual Operation. 100 Illustrations, cloth gilt,
1s. 6d.

Dr. Lardner's School Handbooks.

NATURAL PHILOSOPHY FOR SCHOOLS. By Dr. LARDNER.
328 Illustrations. Sixth Edition. One Vol., 3s. 6d. cloth.

"A very convenient class-book for junior students in private schools. It is intended to convey,
in clear and precise terms, general notions of all the principal divisions of Physical Science."—
British Quarterly Review.

ANIMAL PHYSIOLOGY FOR SCHOOLS. By Dr. LARDNER.
With 190 Illustrations. Second Edition. One Vol., 3s. 6d. cloth.

"Clearly written, well arranged, and excellently illustrated."—*Gardener's Chronicle*.

COUNTING-HOUSE WORK, TABLES, etc.

Introduction to Business.
LESSONS IN COMMERCE. By Professor R. Gambaro, of the Royal High Commercial School at Genoa. Edited and Revised by James Gault, Professor of Commerce and Commercial Law in King's College, London. Crown 8vo, price about 3s. 6d. [*In the press.*

Accounts for Manufacturers.
FACTORY ACCOUNTS: Their Principles and Practice. A Handbook for Accountants and Manufacturers, with Appendices on the Nomenclature of Machine Details; the Income Tax Acts; the Rating of Factories; Fire and Boiler Insurance; the Factory and Workshop Acts, &c., including also a Glossary of Terms and a large number of Specimen Rulings. By Emile Garcke and J. M. Fells. Third Edition. Demy 8vo, 250 pages, price 6s. strongly bound.

"A very interesting description of the requirements of Factory Accounts. . . . the princip.e of assimilating the Factory Accounts to the general commercial books is one which we thoroughly agree with."—*Accountants' Journal.*
"There are few owners of Factories who would not derive great benefit from the perusal of this most admirable work."—*Local Government Chronicle.*

Foreign Commercial Correspondence.
THE FOREIGN COMMERCIAL CORRESPONDENT: Being Aids to Commercial Correspondence in Five Languages—English, French, German, Italian and Spanish. By Conrad E. Baker. Second Edition, Revised. Crown 8vo, 3s. 6d. cloth.

"Whoever wishes to correspond in all the languages mentioned by Mr. Baker cannot do better than study this work, the materials of which are excellent and conveniently arranged."—*Athenæum.*
"A careful examination has convinced us that it is unusually complete, well arranged and reliable. The book is a thoroughly good one."—*Schoolmaster.*

Intuitive Calculations.
THE COMPENDIOUS CALCULATOR; or, Easy and Concise Methods of Performing the various Arithmetical Operations required in Commercial and Business Transactions, together with Useful Tables. By D. O'Gorman. Corrected by Professor J. R. Young. Twenty-seventh Ed., Revised by C. Norris. Fcap. 8vo, 2s. 6d. cloth; or, 3s. 6d. half-bound.

"It would be difficult to exaggerate the usefulness of a book like this to everyone engaged in commerce or manufacturing industry."—*Knowledge.*
"Supplies special and rapid methods for all kinds of calculations. Of great utility to persons engaged in any kind of commercial transactions."—*Scotsman.*

Modern Metrical Units and Systems.
MODERN METROLOGY: A Manual of the Metrical Units and Systems of the Present Century. With an Appendix containing a proposed English System. By Lowis D'A. Jackson, A.M.Inst.C.E., Author of "Aid to Survey Practice," &c. Large crown 8vo, 12s. 6d. cloth.

"The author has brought together much valuable and interesting information. . . . We cannot but recommend the work."—*Nature.*
"For exhaustive tables of equivalent weights and measures of all sorts, and for clear demonstrations of the effects of the various systems that have been proposed or adopted, Mr. Jackson's treatise is without a rival."—*Academy.*

The Metric System and the British Standards.
A SERIES OF METRIC TABLES, in which the British Standard Measures and Weights are compared with those of the Metric System at present in Use on the Continent. By C. H. Dowling, C.E. 8vo, 10s. 6d. strongly bound.

"Their accuracy has been certified by Professor Airy, the Astronomer-Royal."—*Builder.*
"Mr. Dowling's Tables are well put together as a ready-reckoner for the conversion of one system into the other."—*Athenæum.*

Iron and Metal Trades' Calculator.
THE IRON AND METAL TRADES' COMPANION. For expeditiously ascertaining the Value of any Goods bought or sold by Weight, from 1s. per cwt. to 112s. per cwt., and from one farthing per pound to one shilling per pound. Each Table extends from one pound to 100 tons. To which are appended Rules on Decimals, Square and Cube Root, Mensuration of Superficies and Solids, &c.; also Tables of Weights of Materials, and other Useful Memoranda. By Thos. Downie. Strongly bound in leather, 396 pp., 9s.

"A most useful set of tables. . . . Nothing like them before existed."—*Building News.*
"Although specially adapted to the Iron and metal trades, the tables will be found useful in every other business in which merchandise is bought and sold by weight."—*Railway News.*

Calculator for Numbers and Weights Combined.

THE NUMBER, WEIGHT AND FRACTIONAL CALCU-LATOR. Containing upwards of 250,000 Separate Calculations, showing at a glance the value at 422 different rates, ranging from $\frac{1}{16}$th of a Penny to 20s. each, or per cwt., and £20 per ton, of any number of articles consecutively, from 1 to 470.—Any number of cwts., qrs., and lbs., from 1 cwt. to 470 cwts.—Any number of tons, cwts., qrs., and lbs., from 1 to 1,000 tons. By WILLIAM CHADWICK, Public Accountant. Third Edition, Revised and Improved. 8vo, price 18s., strongly bound for Office wear and tear.

*** *This work is specially adapted for the Apportionment of Mileage Charges for Railway Traffic.*

☞ *This comprehensive and entirely unique and original Calculator is adapted for the use of Accountants and Auditors, Railway Companies, Canal Companies, Shippers, Shipping Agents, General Carriers, etc.*

Ironfounders, Brassfounders, Metal Merchants, Iron Manufacturers, Ironmongers, Engineers, Machinists, Boiler Makers, Millwrights, Roofing, Bridge and Girder Makers, Colliery Proprietors, etc.

Timber Merchants, Builders, Contractors, Architects, Surveyors, Auctioneers Valuers, Brokers, Mill Owners and Manufacturers, Mill Furnishers, Merchants and General Wholesale Tradesmen.

*** OPINIONS OF THE PRESS.

"The book contains the answers to questions, and not simply a set of ingenious puzzle methods of arriving at results. It is as easy of reference for any answer or any number of answers as a dictionary, and the references are even more quickly made. For making up accounts or estimates, the book must prove invaluable to all who have any considerable quantity of calculations involving price and measure in any combination to do."—*Engineer.*

"The most perfect work of the kind yet prepared."—*Glasgow Herald.*

Comprehensive Weight Calculator.

THE WEIGHT CALCULATOR. Being a Series of Tables upon a New and Comprehensive Plan, exhibiting at One Reference the exact Value of any Weight from 1 lb. to 15 tons, at 300 Progressive Rates, from 1d. to 168s. per cwt., and containing 186,000 Direct Answers, which, with their Combinations, consisting of a single addition (mostly to be performed at sight), will afford an aggregate of 10,266,000 Answers; the whole being calculated and designed to ensure correctness and promote despatch. By HENRY HARBEN, Accountant. Fourth Edition, carefully Corrected. Royal 8vo, strongly half-bound, £1 5s.

"A practical and useful work of reference for men of business generally ; it is the best of the kind we have seen.' —*Ironmonger.*

"Of priceless value to business men. It is a necessary book in all mercantile offices."—*Sheffield Independent.*

Comprehensive Discount Guide.

THE DISCOUNT GUIDE. Comprising several Series of Tables for the use of Merchants, Manufacturers, Ironmongers, and others, by which may be ascertained the exact Profit arising from any mode of using Discounts, either in the Purchase or Sale of Goods, and the method of either Altering a Rate of Discount or Advancing a Price, so as to produce, by one operation, a sum that will realise any required profit after allowing one or more Discounts: to which are added Tables of Profit or Advance from 1¼ to 90 per cent., Tables of Discount from 1¼ to 98¾ per cent., and Tables of Commission, &c., from ¼ to 10 per cent. By HENRY HARBEN, Accountant, Author of "The Weight Calculator." New Edition, carefully Revised and Corrected. Demy 8vo, 544 pp. half-bound, £1 5s.

"A book such as this can only be appreciated by business men, to whom the saving of time means saving of money. We have the high authority of Professor J. R. Young that the tables throughout the work are constructed upon strictly accurate principles. The work is a model of typographical clearness, and must prove of great value to merchants, manufacturers, and general traders."—*British Trade Journal.*

Iron Shipbuilders' and Merchants' Weight Tables.

IRON-PLATE WEIGHT TABLES: For Iron Shipbuilders, *Engineers and Iron Merchants.* Containing the Calculated Weights of upwards of 150,000 different sizes of Iron Plates, from 1 foot by 6 in. by ¼ in. to 10 feet by 5 feet by 1 in. Worked out on the basis of 40 lbs. to the square foot of Iron of 1 inch in thickness. Carefully compiled and thoroughly Revised by H. BURLINSON and W. H. SIMPSON. Oblong 4to, 25s. half-bound.

"This work will be found of great utility. The authors have had much practical experience of what is wanting in making estimates; and the use of the book will save much time in making elaborate calculations."—*English Mechanic.*

INDUSTRIAL AND USEFUL ARTS.

Soap-making.

THE ART OF SOAP-MAKING : *A Practical Handbook of the Manufacture of Hard and Soft Soaps, Toilet Soaps, etc.* Including many New Processes, and a Chapter on the Recovery of Glycerine from Waste Leys. By ALEXANDER WATT, Author of " Electro-Metallurgy Practically Treated," &c. With numerous Illustrations. Fourth Edition, Revised and Enlarged. Crown 8vo, 7s. 6d. cloth.

"The work will prove very useful, not merely to the technological student, but to the practical soap-boiler who wishes to understand the theory of his art."—*Chemical News.*

"Mr. Watt's book is a thoroughly practical treatise on an art which has almost no literature in our language. We congratulate the author on the success of his endeavour to fill a void in English technical literature."—*Nature.*

Paper Making.

THE ART OF PAPER MAKING : *A Practical Handbook of the Manufacture of Paper from Rags, Esparto, Straw and other Fibrous Materials,* Including the Manufacture of Pulp from Wood Fibre, with a Description of the Machinery and Appliances used. To which are added Details of Processes for Recovering Soda from Waste Liquors. By ALEXANDER WATT. With Illustrations. Crown 8vo, 7s. 6d. cloth.

"This book is succinct, lucid, thoroughly practical, and includes everything of interest to the modern paper maker. It is the latest, most practical and most complete work on the paper-making art before the British public."—*Paper Record.*

" It may be regarded as the standard work on the subject. The book is full of valuable information. The 'Art of Paper-making,' is in every respect a model of a text-book, either for a technical class or for the private student."—*Paper and Printing Trades Journal.*

"Admirably adapted for general as well as ordinary technical reference, and as a handbook for students in technical education may be warmly commended."—*The Paper Maker's Monthly Journal.*

Leather Manufacture.

THE ART OF LEATHER MANUFACTURE. Being a Practical Handbook, in which the Operations of Tanning, Currying, and Leather Dressing are fully Described, the Principles of Tanning Explained and many Recent Processes introduced. By ALEXANDER WATT, Author of " Soap-Making," &c. With numerous Illustrations. Second Edition. Crown 8vo, 9s. cloth.

"A sound, comprehensive treatise on tanning and its accessories. This book is an eminently valuable production, which redounds to the credit of both author and publishers."—*Chemical Review.*

"This volume is technical without being tedious, comprehensive and complete without being prosy, and it bears on every page the impress of a master hand. We have never come across a better trade treatise, nor one that so thoroughly supplied an absolute want."—*Shoe and Leather Trades' Chronicle.*

Boot and Shoe Making.

THE ART OF BOOT AND SHOE-MAKING. A Practical Handbook, including Measurement, Last-Fitting, Cutting-Out, Closing and Making, with a Description of the most approved Machinery employed. By JOHN B. LENO, late Editor of *St. Crispin,* and *The Boot and Shoe-Maker.* With numerous Illustrations. Third Edition. 12mo, 2s. cloth limp.

"This excellent treatise is by far the best work ever written on the subject. A new work, embracing all modern improvements, was much wanted. This want is now satisfied. The chapter on clicking, which shows how waste may be prevented, will save fifty times the price of the book."—*Scottish Leather Trader.*

Dentistry.

MECHANICAL DENTISTRY : *A Practical Treatise on the Construction of the various kinds of Artificial Dentures.* Comprising also Useful Formulæ, Tables and Receipts for Gold Plate, Clasps, Solders, &c. &c. By CHARLES HUNTER. Third Edition, Revised. With upwards of 100 Wood Engravings. Crown 8vo, 3s. 6d. cloth.

" The work is very practical."—*Monthly Review of Dental Surgery.*

" We can strongly recommend Mr. Hunter's treatise to all students preparing for the profession of dentistry, as well as to every mechanical dentist.'—*Dublin Journal of Medical Science.*

Wood Engraving.

WOOD ENGRAVING : *A Practical and Easy Introduction to the Study of the Art.* By WILLIAM NORMAN BROWN. Second Edition. With numerous Illustrations. 12mo, 1s. 6d. cloth limp.

"The book is clear and complete, and will be useful to anyone wanting to understand the first elements of the beautiful art of wood engraving."—*Graphic.*

HANDYBOOKS FOR HANDICRAFTS. By PAUL N. HASLUCK.

Metal Turning.

THE METAL TURNER'S HANDYBOOK. A Practical Manual for Workers at the Foot-Lathe: Embracing Information on the Tools, Appliances and Processes employed in Metal Turning. By PAUL N. HASLUCK, Author of "Lathe-Work." With upwards of One Hundred Illustrations. Second Edition, Revised. Crown 8vo, 2s. cloth.

"Clearly and concisely written, excellent in every way."—*Mechanical World*.

Wood Turning.

THE WOOD TURNER'S HANDYBOOK. A Practical Manual for Workers at the Lathe: Embracing Information on the Tools, Appliances and Processes Employed in Wood Turning. By PAUL N. HASLUCK. With upwards of One Hundred Illustrations. Crown 8vo, 2s. cloth.

"We recommend the book to young turners and amateurs. A multitude of workmen have hitherto sought in vain for a manual of this special industry."—*Mechanical World*.

WOOD AND METAL TURNING. By P. N. HASLUCK. (Being the Two preceding Vols. bound together.) 300 pp, with upwards of 200 Illustrations, crown 8vo, 3s. 6d. cloth.

Watch Repairing.

THE WATCH JOBBER'S HANDYBOOK. A Practical Manual on Cleaning, Repairing and Adjusting. Embracing Information on the Tools, Materials, Appliances and Processes Employed in Watchwork. By PAUL N. HASLUCK. With upwards of One Hundred Illustrations. Cr. 8vo, 2s. cloth.

"All young persons connected with the trade should acquire and study this excellent, and at the same time, inexpensive work."—*Clerkenwell Chronicle*.

Clock Repairing.

THE CLOCK JOBBER'S HANDYBOOK : A Practical Manual on Cleaning, Repairing and Adjusting. Embracing Information on the Tools, Materials, Appliances and Processes Employed in Clockwork. By PAUL N. HASLUCK. With upwards of 100 Illustrations. Cr. 8vo, 2s. cloth.

"Of inestimable service to those commencing the trade."—*Coventry Standard*.

WATCH AND CLOCK JOBBING. By P. N. HASLUCK. (Being the Two preceding Vols. bound together.) 320 pp., with upwards of 200 Illustrations, crown 8vo, 3s. 6d. cloth.

Pattern Making.

THE PATTERN MAKER'S HANDYBOOK. A Practical Manual, embracing Information on the Tools, Materials and Appliances employed in Constructing Patterns for Founders. By PAUL N. HASLUCK. With One Hundred Illustrations. Crown 8vo, 2s. cloth.

"This handy volume contains sound information of considerable value to students and artificers."—*Hardware Trades Journal*.

Mechanical Manipulation.

THE MECHANIC'S WORKSHOP HANDYBOOK. A Practical Manual on Mechanical Manipulation. Embracing Information on various Handicraft Processes, with Useful Notes and Miscellaneous Memoranda. By PAUL N. HASLUCK. Crown 8vo, 2s. cloth.

"It is a book which should be found in every workshop, as it is one which will be continually referred to for a very great amount of standard information."—*Saturday Review*.

Model Engineering.

THE MODEL ENGINEER'S HANDYBOOK : A Practical Manual on Model Steam Engines. Embracing Information on the Tools, Materials and Processes Employed in their Construction. By PAUL N. HASLUCK. With upwards of 100 Illustrations. Crown 8vo, 2s. cloth.

"By carefully going through the work, amateurs may pick up an excellent notion of the construction of full-sized steam engines."—*Telegraphic Journal*.

Cabinet Making.

THE CABINET WORKER'S HANDYBOOK : A Practical Manual, embracing Information on the Tools, Materials, Appliances and Processes employed in Cabinet Work. By PAUL N. HASLUCK, Author of "Lathe Work," &c. With upwards of 100 Illustrations. Crown 8vo, 2s. cloth.

[Glasgow Herald.

"Thoroughly practical throughout. The amateur worker in wood will find it most useful."—

Electrolysis of Gold, Silver, Copper, etc.

ELECTRO-DEPOSITION : A Practical Treatise on the Electrolysis of Gold, Silver, Copper, Nickel, and other Metals and Alloys. With descriptions of Voltaic Batteries, Magneto and Dynamo-Electric Machines, Thermopiles, and of the Materials and Processes used in every Department of the Art, and several Chapters on Electro-Metallurgy. By ALEXANDER WATT. Third Edition, Revised and Corrected. Crown 8vo, 9s. cloth.
"Eminently a book for the practical worker in electro-deposition. It contains practical descriptions of methods, processes and materials as actually pursued and used in the workshop."
—*Engineer.*

Electro-Metallurgy.

ELECTRO-METALLURGY ; Practically Treated. By ALEXANDER WATT. Author of "Electro-Deposition," &c. Ninth Edition, Enlarged and Revised, with Additional Illustrations, and including the most recent Processes. 12mo, 4s. cloth boards.
"From this book both amateur and artisan may learn everything necessary for the successfu prosecution of electroplating."—*Iron.*

Electroplating.

ELECTROPLATING : A Practical Handbook on the Deposition of Copper, Silver, Nickel, Gold, Aluminium, Brass, Platinum, &c. &c. With Descriptions of the Chemicals, Materials, Batteries and Dynamo Machines used in the Art. By J. W. URQUHART, C.E. Second Edition, with Additions. Numerous Illustrations. Crown 8vo, 5s. cloth.
" An excellent practical manual."—*Engineering.*
" An excellent work, giving the newest information."—*Horological Journal.*

Electrotyping.

ELECTROTYPING : The Reproduction and Multiplication of Printing Surfaces and Works of Art by the Electro-deposition of Metals. By J. W. URQUHART, C.E. Crown 8vo, 5s. cloth.
'The book is thoroughly practical. The reader is, therefore, conducted through the leading aws of electricity, then through the metals used by electrotypers, the apparatus, and the depositing processes, up to the final preparation of the work."—*Art Journal.*

Horology.

A TREATISE ON MODERN HOROLOGY, in Theory and Practice. Translated from the French of CLAUDIUS SAUNIER, by JULIEN TRIPPLIN, F.R.A.S., and EDWARD RIGG, M.A., Assayer in the Royal Mint. With 78 Woodcuts and 22 Coloured Plates. Second Edition. Royal 8vo, £2 2s. cloth ; £2 10s. half-calf.
" There is no horological work in the English language at all to be compared to this production of M. Saunier's for clearness and completeness. It is alike good as a guide for the student and as a reference for the experienced horologist and skilled workman."—*Horological Journal.*
" The latest, the most complete, and the most reliable of those literary productions to which continental watchmakers are indebted for the mechanical superiority over their English brethren —in fact, the Book of Books, is M. Saunier's 'Treatise.'"—*Watchmaker, Jeweller and Silversmith.*

Watchmaking.

THE WATCHMAKER'S HANDBOOK. A Workshop Companion for those engaged in Watchmaking and the Allied Mechanical Arts. From the French of CLAUDIUS SAUNIER. Enlarged by JULIEN TRIPPLIN, F.R.A.S., and EDWARD RIGG, M.A., Assayer in the Royal Mint. Woodcuts and Copper Plates. Third Edition, Revised. Crown 8vo, 9s. cloth.
" Each part is truly a treatise in itself. The arrangement is good and the language is clear and concise. It is an admirable guide for the young watchmaker."—*Engineering.*
" It is impossible to speak too highly of its excellence. It fulfils every requirement in a handbook intended for the use of a workman."—*Watch and Clockmaker.*
" This book contains an immense number of practical details bearing on the daily occupation of a watchmaker."—*Watchmaker and Metalworker* (Chicago).

Goldsmiths' Work.

THE GOLDSMITH'S HANDBOOK. By GEORGE E. GEE, Jeweller, &c. Third Edition, considerably Enlarged. 12mo, 3s. 6d. cl. bds.
"A good, sound educator, and will be accepted as an authority."—*Horological Journal.*

Silversmiths' Work.

THE SILVERSMITH'S HANDBOOK. By GEORGE E. GEE, Jeweller, &c. Second Edition, Revised, with numerous Illustrations. 12mo, 3s. 6d. cloth boards.
"Workers in the trade will speedily discover its merits when they sit down to study it."—*English Mechanic.*
** The above two works together, strongly half-bound, price 7s.

Bread and Biscuit Baking.

THE BREAD AND BISCUIT BAKER'S AND SUGAR-BOILER'S ASSISTANT. Including a large variety of Modern Recipes. With Remarks on the Art of Bread-making. By ROBERT WELLS, Practical Baker. Second Edition, with Additional Recipes. Crown 8vo, 2s. cloth.

" A large number of wrinkles for the ordinary cook, as well as the baker."—*Saturday Review.*

Confectionery.

THE PASTRYCOOK AND CONFECTIONER'S GUIDE. For Hotels, Restaurants and the Trade in general, adapted also for Family Use. By ROBERT WELLS, Author of " The Bread and Biscuit Baker's and Sugar Boiler's Assistant." Crown 8vo, 2s. cloth.

" We cannot speak too highly of this really excellent work. In these days of keen competition our readers cannot do better than purchase this book."—*Bakers' Times.*

Ornamental Confectionery.

ORNAMENTAL CONFECTIONERY: A Guide for Bakers. Confectioners and Pastrycooks; including a variety of Modern Recipes, and Remarks on Decorative and Coloured Work. With 129 Original Designs. By ROBERT WELLS. Crown 8vo, 5s. cloth.

" A valuable work, and should be in the hands of every baker and confectioner. The illustrative designs are alone worth treble the amount charged for the whole work."—*Bakers' Times.*

Flour Confectionery.

THE MODERN FLOUR CONFECTIONER. Wholesale and Retail. Containing a large Collection of Recipes for Cheap Cakes, Biscuits, &c. With Remarks on the Ingredients used in their Manufacture, &c. By R. WELLS, Author of "Ornamental Confectionery," "The Bread and Biscuit Baker," "The Pastrycook's Guide," &c. Crown 8vo, 2s. cloth.

Laundry Work.

LAUNDRY MANAGEMENT. A Handbook for Use in Private and Public Laundries, Including Descriptive Accounts of Modern Machinery and Appliances for Laundry Work. By the EDITOR of " The Laundry Journal." With numerous Illustrations. Crown 8vo, 2s. 6d. cloth.

CHEMICAL MANUFACTURES & COMMERCE.

New Manual of Engineering Chemistry.

ENGINEERING CHEMISTRY: A Practical Treatise for the Use of Analytical Chemists, Engineers, Iron Masters, Iron Founders, Students, and others. Comprising Methods of Analysis and Valuation of the Principal Materials used in Engineering Work, with numerous Analyses, Examples, and Suggestions. By H. JOSHUA PHILLIPS, F.I.C., F.C.S., Analytical and Consulting Chemist to the Great Eastern Railway. Crown 8vo, 320 pp., with Illustrations, 10s. 6d. cloth. [*Just published.*

" In this work the author has rendered no small service to a numerous body of practical men. . . . The analytical methods may be pronounced most satisfactory, being as accurate as the despatch required of engineering chemists permits."—*Chemical News.*

Analysis and Valuation of Fuels.

FUELS: SOLID, LIQUID AND GASEOUS, Their Analysis and Valuation. For the Use of Chemists and Engineers. By H. J. PHILLIPS, F.C.S., Analytical and Consulting Chemist to the Great Eastern Railway. Crown 8vo, 3s. 6d. cloth.

" Ought to have its place in the laboratory of every metallurgical establishment, and wherever fuel is used on a large scale."—*Chemical News.*

" Cannot fail to be of wide interest, especially at the present time."—*Railway News.*

Alkali Trade, Manufacture of Sulphuric Acid, etc.

A MANUAL OF THE ALKALI TRADE, including the Manufacture of Sulphuric Acid, Sulphate of Soda, and Bleaching Powder. By JOHN LOMAS. 390 pages. With 232 Illustrations and Working Drawings. Second Edition. Royal 8vo, £1 10s. cloth.

" This book is written by a manufacturer for manufacturers. The working details of the most approved forms of apparatus are given, and these are accompanied by no less than 232 wood engravings, all of which may be used for the purposes of construction."—*Athenæum.*

The Blowpipe.

THE BLOWPIPE IN CHEMISTRY, MINERALOGY, AND GEOLOGY. Containing all known Methods of Anhydrous Analysis, Working Examples, and Instructions for Making Apparatus. By Lieut.-Col. W. A. Ross, R.A. With 120 Illustrations. New Edition. Crown 8vo, 5s.

"The student who goes through the course of experimentation here laid down will gain a better insight into inorganic chemistry and mineralogy than if he had 'got up' any of the best text-books of the day, and passed any number of examinations in their contents."—*Chemical News.*

Commercial Chemical Analysis.

THE COMMERCIAL HANDBOOK OF CHEMICAL ANALYSIS; or, Practical Instructions for the determination of the Intrinsic or Commercial Value of Substances used in Manufactures, Trades, and the Arts. By A. NORMANDY. New Edition by H. M. NOAD, F.R.S. Cr. 8vo, 12s. 6d. cl.

"Essential to the analysts appointed under the new Act. The most recent results are given, and the work is well edited and carefully written."—*Nature.*

Brewing.

A HANDBOOK FOR YOUNG BREWERS. By HERBERT EDWARDS WRIGHT, B.A. New Edition, much Enlarged. [*In the press.*

Dye-Wares and Colours.

THE MANUAL OF COLOURS AND DYE-WARES : Their Properties, Applications, Valuation, Impurities, and Sophistications. For the use of Dyers, Printers, Drysalters, Brokers, &c. By J. W. SLATER. Second Edition. Revised and greatly Enlarged. Crown 8vo, 7s. 6d. cloth.

"A complete encyclopædia of the *materia tinctoria*. The information given respecting each article is full and precise, and the methods of determining the value of articles such as these, so liable to sophistication, are given with clearness, and are practical as well as valuable."—*Chemist and Druggist.*

"There is no other work which covers precisely the same ground. To students preparing or examinations in dyeing and printing it will prove exceedingly useful."—*Chemical News.*

Pigments.

THE ARTIST'S MANUAL OF PIGMENTS. Showing their Composition, Conditions of Permanency. Non-Permanency, and Adulterations : Effects in Combination with Each Other and with Vehicles ; and the most Reliable Tests of Purity. By H. C. STANDAGE Second Edition. Crown 8vo, 2s. 6d. cloth.

"This work is indeed *multum-in-parvo*, and we can, with good conscience, recommend it to all who come in contact with pigments, whether as makers, dealers or users."—*Chemical Review.*

Gauging. Tables and Rules for Revenue Officers, Brewers, etc.

A POCKET BOOK OF MENSURATION AND GAUGING : Containing Tables, Rules and Memoranda for Revenue Officers, Brewers, Spirit Merchants, &c. By J. B. MANT (Inland Revenue). Second Edition Revised. Oblong 18mo, 4s. leather, with elastic band.

"This handy and useful book is adapted to the requirements of the Inland Revenue Department, and will be a favourite book of reference."—*Civilian.*

"Should be in the hands of every practical brewer."—*Brewers' Journal.*

AGRICULTURE, FARMING, GARDENING, etc.

Youatt and Burn's Complete Grazier.

THE COMPLETE GRAZIER, and FARMER'S and CATTLE-BREEDER'S ASSISTANT. Including the Breeding, Rearing, and Feeding of Stock ; Management of the Dairy, Culture and Management of Grass Land, and of Grain and Root Crops, &c. By W. YOUATT and R. SCOTT BURN. An entirely New Edition, partly Re-written and greatly Enlarged, by W. FREAM, B.Sc.Lond., LL.D. In medium 8vo, about 1,000 pp. [*In the press.*

Agricultural Facts and Figures.

NOTE-BOOK OF AGRICULTURAL FACTS AND FIGURES FOR FARMERS AND FARM STUDENTS. By PRIMROSE McCONNELL. late Professor of Agriculture, Glasgow Veterinary College. Third Edition. Royal 32mo, 4s. leather.

"The most complete and comprehensive Note-book for Farmers and Farm Students that we have seen. It literally teems with information, and we can cordially recommend it to all connected with agriculture."—*North British Agriculturist.*

Flour Manufacture, Milling, etc.

FLOUR MANUFACTURE: A Treatise on Milling Science
and Practice. By FRIEDRICH KICK, Imperial Regierungsrath, Professor of
Mechanical Technology in the Imperial German Polytechnic Institute,
Prague. Translated from the Second Enlarged and Revised Edition with
Supplement. By H. H. P. POWLES, A.M.I.C.E. Nearly 400 pp. Illustrated
with 28 Folding Plates, and 167 Woodcuts. Royal 8vo, 25s. cloth.

" This valuable work is, and will remain, the standard authority on the science of milling. . . .
The miller who has read and digested this work will have laid the foundation, so to speak, of a successful career ; he will have acquired a number of general principles which he can proceed to
apply.   In this handsome volume we at last have the accepted text-book of modern milling in good,
sound English, which has little, if any, trace of the German idiom."—*The Miller.*
" The appearance of this celebrated work in English is very opportune, and British millers
will, we are sure, not be slow in availing themselves of its pages."—*Millers' Gazette.*

Small Farming.

SYSTEMATIC SMALL FARMING; *or, The Lessons of my
Farm.* Being an Introduction to Modern Farm Practice for Small Farmers
in the Culture of Crops; The Feeding of Cattle; The Management of the
Dairy, Poultry and Pigs, &c. &c.   By ROBERT SCOTT BURN, Author of "Outlines of Landed Estates' Management." Numerous Illusts., cr. 8vo, 6s. cloth.

" This is the completest book of its class we have seen, and one which every amateur farmer
will read with pleasure and accept as a guide."—*Field.*
" The volume contains a vast amount of useful information.   No branch of farming is left
untouched, from the labour to be done to the results achieved.   It may be safely recommended to
all who think they will be in paradise when they buy or rent a three-acre farm."—*Glasgow Herald*

Modern Farming.

OUTLINES OF MODERN FARMING. By R. SCOTT BURN.
Soils, Manures, and Crops—Farming and Farming Economy—Cattle, Sheep,
and Horses — Management of Dairy, Pigs and Poultry — Utilisation of
Town-Sewage, Irrigation, &c.   Sixth Edition.   In One Vol., 1,250 pp., half-
bound, profusely Illustrated, 12s.

" The aim of the author has been to make his work at once comprehensive and trustworthy,
and in this aim he has succeeded to a degree which entitles him to much credit."—*Morning
Advertiser.*   " No farmer should be without this book."—*Banbury Guardian.*

Agricultural Engineering.

FARM ENGINEERING, THE COMPLETE TEXT-BOOK OF.
Comprising Draining and Embanking; Irrigation and Water Supply ; Farm
Roads, Fences, and Gates; Farm Buildings, their Arrangement and Con-
struction, with Plans and Estimates; Barn Implements and Machines; Field
Implements and Machines; Agricultural Surveying, Levelling, &c.   By Prof.
JOHN SCOTT, Editor of the " Farmers' Gazette," late Professor of Agriculture
and Rural Economy at the Royal Agricultural College, Cirencester, &c. &c.
In One Vol., 1,150 pages, half-bound, with over 600 Illustrations, 12s.

" Written with great care, as well as with knowledge and ability.   The author has done his
work well ; we have found him a very trustworthy guide wherever we have tested his statements.
The volume will be of great value to agricultural students."—*Mark Lane Express.*
" For a young agriculturist we know of no handy volume likely to be more usefully studied.
—*Bell's Weekly Messenger.*

English Agriculture.

THE FIELDS OF GREAT BRITAIN : A Text-Book of
Agriculture, adapted to the Syllabus of the Science and Art Department.
For Elementary and Advanced Students. By HUGH CLEMENTS (Board of
Trade). Second Ed., Revised, with Additions. 18mo, 2s. 6d. cl.

" A most comprehensive volume, giving a mass of information."—*Agricultural Economist.*
" It is a long time since we have seen a book which has pleased us more, or which contains
such a vast and useful fund of knowledge."—*Educational Times.*

Tables for Farmers, etc.

TABLES, MEMORANDA, AND CALCULATED RESULTS
*for Farmers, Graziers, Agricultural Students, Surveyors, Land Agents Auc-
tioneers, etc.*   With a New System of Farm Book-keeping.   Selected and
Arranged by SIDNEY FRANCIS.   Second Edition, Revised.   272 pp., waistcoat-pocket size, 1s. 6d. limp leather.

" Weighing less than 1 oz., and occupying no more space than a match box, it contains a mass
of facts and calculations which has never before, in such handy form, been obtainable. . Every
operation on the farm is dealt with.   The work may be taken as thoroughly accurate, the whole of
the tables having been revised by Dr. Fream.   We cordially recommend it."—*Bell's Weekly
Messenger.*
" A marvellous little book. . . .   The agriculturist who possesses himself of it will not be
disappointed with his investment."—*The Farm.*

Farm and Estate Book-keeping.

BOOK-KEEPING FOR FARMERS & ESTATE OWNERS.
A Practical Treatise, presenting, in Three Plans, a System adapted for all
Classes of Farms. By JOHNSON M. WOODMAN, Chartered Accountant. Second
Edition, Revised. Cr. 8vo, 3s. 6d. cl. bds. ; or 2s. 6d. cl. limp.
" The volume is a capital study of a most important subject." *Agricultural Gazette.*
" Will be found of great assistance by those who intend to commence a system of book-keep-
ing, the author's examples being clear and explicit, and his explanations, while full and accurate,
eing to a large extent free from technicalities."—*Live Stock Journal.*

Farm Account Book.

WOODMAN'S YEARLY FARM ACCOUNT BOOK. Giving
a Weekly Labour Account and Diary, and showing the Income and Expen-
diture under each Department of Crops, Live Stock, Dairy, &c. &c. With
Valuation, Profit and Loss Account, and Balance Sheet at the end of the
Year, and an Appendix of Forms. Ruled and Headed for Entering a Com-
plete Record of the Farming Operations. By JOHNSON M. WOODMAN,
Chartered Accountant. Folio, 7s. 6d. half bound.   [*culture.*
"Contains every requisite form for keeping farm accounts readily and accurately."—*Agri*

Early Fruits, Flowers and Vegetables.

THE FORCING GARDEN ; or, How to Grow Early Fruits,
Flowers, and Vegetables. With Plans and Estimates for Building Glass-
houses, Pits and Frames. By SAMUEL WOOD. Crown 8vo, 3s. 6d. cloth.
"A good book, and fairly fills a place that was in some degree vacant. The book is written with
great care, and contains a great deal of valuable teaching."—*Gardeners' Magazine.*
" Mr. Wood's book is an original and exhaustive answer to the question ' How to Grow Early
Fruits, Flowers and Vegetables ? '"—*Land and Water.*

Good Gardening.

A PLAIN GUIDE TO GOOD GARDENING ; or, How to Grow
Vegetables, Fruits, and Flowers. With Practical Notes on Soils, Manures,
Seeds, Planting, Laying-out of Gardens and Grounds, &c. By S. WOOD.
Fourth Edition, with numerous Illustrations. Crown 8vo, 3s. 6d. cloth.
"A very good book, and one to be highly recommended as a practical guide. The practical
directions are excellent."—*Athenæum.*
" May be recommended to young gardeners, cottagers, and specially to amateurs, for the
plain, simple, and trustworthy information it gives on common matters too often neglected."—
Gardeners' Chronicle.

Gainful Gardening.

MULTUM-IN-PARVO GARDENING ; or, How to make One
Acre of Land produce £620 a-year by the Cultivation of Fruits and Vegetables ;
also, How to Grow Flowers in Three Glass Houses, so as to realise £176 per
annum clear Profit. By S. WOOD. Fifth Edition. Crown 8vo, 1s. sewed.
" We are bound to recommend it as not only suited to the case of the amateur and gentleman's
gardener, but to the market grower."—*Gardeners' Magazine.*

Gardening for Ladies.

THE LADIES' MULTUM-IN-PARVO FLOWER GARDEN,
and *Amateurs' Complete Guide.* By S. WOOD. With Illusts. Cr. 8vo, 3s. 6d. cl.
" This volume contains a good deal of sound, common sense instruction."—*Florist.*
" Full of shrewd hints and useful instructions, based on a lifetime of experience."—*Scotsman.*

Receipts for Gardeners.

GARDEN RECEIPTS. By C. W. QUIN. 12mo, 1s. 6d. cloth.
"A useful and handy book, containing a good deal of valuable information."—*Athenæum.*

Market Gardening.

MARKET AND KITCHEN GARDENING. By Contributors
to "The Garden." Compiled by C. W. SHAW, late Editor of " Gardening
Illustrated." 12mo, 3s. 6d. cloth boards.
" The most valuable compendium of kitchen and market-garden work published."—*Farmer.*

Cottage Gardening.

COTTAGE GARDENING ; *or, Flowers, Fruits, and Vegetables for
Small Gardens.* By E. HOBDAY. 12mo, 1s. 6d. cloth limp.

Potato Culture.

POTATOES : *How to Grow and Show Them.* A Practical Guide
to the Cultivation and General Treatment of the Potato. By JAMES PINK.
Second Edition. Crown 8vo, 2s. cloth.

LAND AND ESTATE MANAGEMENT, LAW, etc.

Hudson's Land Valuer's Pocket-Book.

THE LAND VALUER'S BEST ASSISTANT: Being Tables on a very much Improved Plan, for Calculating the Value of Estates. With Tables for reducing Scotch, Irish, and Provincial Customary Acres to Statute Measure, &c. By R. HUDSON, C.E. New Edition. Royal 32mo, leather, elastic band, 4s.

"This new edition includes tables for ascertaining the value of leases for any term of years; and for showing how to lay out plots of ground of certain acres in forms, square, round, &c., with valuable rules for ascertaining the probable worth of standing timber to any amount; and is of incalculable value to the country gentleman and professional man."—*Farmers' Journal.*

Ewart's Land Improver's Pocket-Book.

THE LAND IMPROVER'S POCKET-BOOK OF FORMULÆ, TABLES and MEMORANDA required in any Computation relating to the Permanent Improvement of Landed Property. By JOHN EWART, Land Surveyor and Agricultural Engineer. Second Edition, Revised. Royal 32mo, oblong, leather, gilt edges, with elastic band, 4s.

"A compendious and handy little volume."—*Spectator.*

Complete Agricultural Surveyor's Pocket-Book.

THE LAND VALUER'S AND LAND IMPROVER'S COM-PLETE POCKET-BOOK. Consisting of the above Two Works bound together. Leather, gilt edges, with strap, 7s. 6d.

"Hudson's book is the best ready-reckoner on matters relating to the valuation of land and crops, and its combination with Mr. Ewart's work greatly enhances the value and usefulness of the latter-mentioned. . . . It is most useful as a manual for reference."—*North of England Farmer.*

Auctioneer's Assistant.

THE APPRAISER, AUCTIONEER, BROKER, HOUSE AND ESTATE AGENT AND VALUER'S POCKET ASSISTANT, for the Valuation for Purchase, Sale, or Renewal of Leases, Annuities and Reversions, and of property generally; with Prices for Inventories, &c. By JOHN WHEELER, Valuer, &c. Fifth Edition, re-written and greatly extended by C. NORRIS, Surveyor, Valuer, &c. Royal 32mo, 5s. cloth.

"A neat and concise book of reference, containing an admirable and clearly-arranged list of prices for inventories, and a very practical guide to determine the value of furniture, &c."—*Standard.*

"Contains a large quantity of varied and useful information as to the valuation for purchase, sale, or renewal of leases, annuities and reversions, and of property generally, with prices for Inventories, and a guide to determine the value of interior fittings and other effects."—*Builder.*

Auctioneering.

AUCTIONEERS: THEIR DUTIES AND LIABILITIES. A Manual of Instruction and Counsel for the Young Auctioneer. By ROBERT SQUIBBS, Auctioneer. Second Edition, Revised and partly Re-written. Demy 8vo, 12s. 6d. cloth.

"The position and duties of auctioneers treated compendiously and clearly."—*Builder.*

"Every auctioneer ought to possess a copy of this excellent work."—*Ironmonger.*

"Of great value to the profession. . . . We readily welcome this book from the fact that it treats the subject in a manner somewhat new to the profession."—*Estates Gazette.*

Legal Guide for Pawnbrokers.

THE PAWNBROKERS', FACTORS' AND MERCHANTS' GUIDE TO THE LAW OF LOANS AND PLEDGES. With the Statutes and a Digest of Cases on Rights and Liabilities, Civil and Criminal, as to Loans and Pledges of Goods, Debentures, Mercantile and other Securities. By H. C. FOLKARD, Esq., Barrister-at-Law, Author of "The Law of Slander and Libel," &c. With Additions and Corrections. Fcap. 8vo, 3s. 6d. cloth.

"This work contains simply everything that requires to be known concerning the department of the law of which it treats. We can safely commend the book as unique and very nearly perfect."—*Iron.*

"The task undertaken by Mr. Folkard has been very satisfactorily performed. . . . Such explanations as are needful have been supplied with great clearness and with due regard to brevity."—*City Press.*

Law of Patents.

PATENTS FOR INVENTIONS, AND HOW TO PROCURE THEM. Compiled for the Use of Inventors, Patentees and others. By G. G. M. HARDINGHAM, Assoc.Mem.Inst.C.E., &c. Demy 8vo, cloth, price 2s. 6d.

Metropolitan Rating Appeals.

REPORTS OF APPEALS HEARD BEFORE THE COURT OF GENERAL ASSESSMENT SESSIONS, from the Year 1871 to 1885. By EDWARD RYDE and ARTHUR LYON RYDE. Fourth Edition, brought down to the Present Date, with an Introduction to the Valuation (Metropolis) Act, 1869, and an Appendix by WALTER C. RYDE, of the Inner Temple, Barrister-at-Law. 8vo, 16s. cloth.

" A useful work, occupying a place mid-way between a handbook for a lawyer and a guide to the surveyor. It is compiled by a gentleman eminent in his profession as a land agent, whose specialty, it is acknowledged, lies i the direction of assessing property for rating purposes."—*Land Agents' Record.*

" It is an indispensable work of reference for all engaged in assessment business."—*Journal of Gas Lighting.*

House Property.

HANDBOOK OF HOUSE PROPERTY. A Popular and Practical Guide to the Purchase, Mortgage, Tenancy, and Compulsory Sale of Houses and Land, including the Law of Dilapidations and Fixtures; with Examples of all kinds of Valuations, Useful Information on Building, and Suggestive Elucidations of Fine Art. By E. L. TARBUCK, Architect and Surveyor. Fourth Edition, Enlarged. 12mo, 5s. cloth.

" The advice is thoroughly practical."—*Law Journal.*
' For all who have dealings with house property, this is an indispensable guide."—*Decoration.*
" Carefully brought up to date, and much improved by the addition of a division on fine art.
" A well-written and thoughtful work."—*Land Agent's Record.*

Inwood's Estate Tables.

TABLES FOR THE PURCHASING OF ESTATES, Freehold, Copyhold, or Leasehold; Annuities, Advowsons, etc., and for the Renewing of Leases held under Cathedral Churches, Colleges, or other Corporate bodies, for Terms of Years certain, and for Lives; also for Valuing Reversionary Estates, Deferred Annuities, Next Presentations, &c.; together with SMART'S Five Tables of Compound Interest, and an Extension of the same to Lower and Intermediate Rates. By W. INWOOD. 23rd Edition, with considerable Additions, and new and valuable Tables of Logarithms for the more Difficult Computations of the Interest of Money, Discount, Annuities, &c., by M. FEDOR THOMAN, of the Société Crédit Mobilier of Paris. Crown 8vo, 8s. cloth.

" Those interested in the purchase and sale of estates, and in the adjustment of compensation cases, as well as in transactions in annuities, life insurances, &c., will find the present edition of eminent service."—*Engineering.*

" 'Inwood's Tables' still maintain a most enviable reputation. The new issue has been enriched by large additional contributions by M. Fedor Thoman, whose carefully arranged Tables cannot fail to be of the utmost utility."—*Mining Journal.*

Agricultural and Tenant-Right Valuation.

THE AGRICULTURAL AND TENANT-RIGHT-VALUER'S ASSISTANT. A Practical Handbook on Measuring and Estimating the Contents, Weights and Values of Agricultural Produce and Timber, the Values of Estates and Agricultural Labour, Forms of Tenant-Right-Valuations, Scales of Compensation under the Agricultural Holdings Act, 1883, &c. &c. By TOM BRIGHT, Agricultural Surveyor. Crown 8vo, 3s. 6d. cloth.

" Full of tables and examples in connection with the valuation of tenant-right, estates, labour, contents, and weights of timber, and farm produce of all kinds."—*Agricultural Gazette.*
" An eminently practical handbook, full of practical tables and data of undoubted interest and value to surveyors and auctioneers in preparing valuations of all kinds."—*Farmer.*

Plantations and Underwoods.

POLE PLANTATIONS AND UNDERWOODS: A Practical Handbook on Estimating the Cost of Forming, Renovating, Improving and Grubbing Plantations and Underwoods, their Valuation for Purposes of Transfer, Rental, Sale or Assessment. By TOM BRIGHT, F.S.Sc., Author of "The Agricultural and Tenant-Right-Valuer's Assistant," &c. Crown 8vo, 3s. 6d. cloth. [*Just published.*
" Will be found very useful to those who are actually engaged in managing wood."—*Bell's Weekly Messenger.*
" To valuers, foresters and agents it will be a welcome aid." *North British Agriculturist.*
"Well calculated to assist the valuer in the discharge of his duties, and of undoubted interest and will both to surveyors and auctioneers in preparing valuations of all kinds." *Kent Herald.*

A Complete Epitome of the Laws of this Country.

EVERY MAN'S OWN LAWYER: A Handy-Book of the Principles of Law and Equity. By A BARRISTER. Twenty-eighth Edition. Revised and Enlarged. Including the Legislation of 1890, and including careful digests of *The Bankruptcy Act*, 1890; the *Directors' Liability Act*, 1890; the *Partnership Act*, 1890; the *Intestates' Estates Act*, 1890; the *Settled Land Act*, 1890; the *Housing of the Working Classes Act*, 1890; the *Infectious Disease (Prevention) Act*, 1890; the *Allotments Act*, 1890; the *Tenants' Compensation Act*, 1890; and the *Trustees' Appointment Act*, 1890; while other new Acts have been duly noted. Crown 8vo, 688 pp., price 6s. 8d. (saved at every consultation!), strongly bound in cloth.   [*Just published.*

*** THE BOOK WILL BE FOUND TO COMPRISE (AMONGST OTHER MATTER)—

THE RIGHTS AND WRONGS OF INDIVIDUALS—LANDLORD AND TENANT—VENDORS AND PURCHASERS—PARTNERS AND AGENTS—COMPANIES AND ASSOCIATIONS—MASTERS, SERVANTS AND WORKMEN—LEASES AND MORTGAGES—CHURCH AND CLERGY, RITUAL —LIBEL AND SLANDER—CONTRACTS AND AGREEMENTS—BONDS AND BILLS OF SALE—CHEQUES, BILLS AND NOTES- RAILWAY AND SHIPPING LAW—BANKRUPTCY AND INSURANCE—BORROWERS, LENDERS AND SURETIES—CRIMINAL LAW—PARLIAMENTARY ELECTIONS—COUNTY COUNCILS—MUNICIPAL CORPORATIONS—PARISH LAW, CHURCHWARDENS, ETC.—INSANITARY DWELLINGS AND AREAS—PUBLIC HEALTH AND NUISANCES—FRIENDLY AND BUILDING SOCIETIES—COPYRIGHT AND PATENTS—TRADE MARKS AND DESIGNS—HUSBAND AND WIFE, DIVORCE, ETC.—TRUSTEES AND EXECUTORS—GUARDIAN AND WARD, INFANTS, ETC.—GAME LAWS AND SPORTING—HORSES, HORSE-DEALING AND DOGS—INNKEEPERS, LICENSING, ETC.—FORMS OF WILLS AGREEMENTS, ETC. ETC.

NOTE.—*The object of this work is to enable those who consult it to help themselves to the law; and thereby to dispense, as far as possible, with professional assistance and advice. There are many wrongs and grievances which persons submit to from time to time through not knowing how or where to apply for redress; and many persons have as great a dread of a lawyer's office as of a lion's den. With this book at hand it is believed that many a SIX-AND-EIGHTPENCE may be saved; many a wrong redressed; many a right reclaimed; many a law suit avoided; and many an evil abated. The work has established itself as the standard legal adviser of all classes, and also made a reputation for itself as a useful book of reference for lawyers residing at a distance from law libraries, who are glad to have at hand a work embodying recent decisions and enactments.*

*** OPINIONS OF THE PRESS.

" It is a complete code of English Law, written in plain language, which all can understand. . . Should be in the hands of every business man, and all who wish to abolish lawyers' bills."—*Weekly Times.*

" A useful and concise epitome of the law, compiled with considerable care."—*Law Magazine.*

"A complete digest of the most useful facts which constitute English law."—*Globe.*

" This excellent handbook. . . . Admirably done, admirably arranged, and admirably cheap."—*Leeds Mercury.*

' A concise, cheap and complete epitome of the English law. So plainly written that he who runs may read, and he who reads may understand."—*Figaro.*

" A dictionary of legal facts well put together. The book is a very useful one."—*Spectator.*

" A work which has long been wanted, which is thoroughly well done, and which we most cordially recommend."—*Sunday Times.*

"The latest edition of this popular book ought to be in every business establishment, and on every library table."—*Sheffield Post.*

Private Bill Legislation and Provisional Orders.

HANDBOOK FOR THE USE OF SOLICITORS AND ENGINEERS Engaged in Promoting Private Acts of Parliament and Provisional Orders, for the Authorization of Railways, Tramways, Works for the Supply of Gas and Water, and other undertakings of a like character. By L. LIVINGSTON MACASSEY, of the Middle Temple, Barrister-at-Law, and Member of the Institution of Civil Engineers; Author of "Hints on Water Supply." Demy 8vo, 950 pp., price 25s. cloth.

"The volume is a desideratum on a subject which can be only acquired by practical experience, and the order of procedure in Private Bill Legislation and Provisional Orders is followed. The author's suggestions and notes will be found of great value to engineers and others professionally engaged in this class of practice."—*Building News.*

" The author's double experience as an engineer and barrister has eminently qualified him for the task, and enabled him to approach the subject alike from an engineering and legal point of view. The volume will be found a great help both to engineers and lawyers engaged in promoting Private Acts of Parliament and Provisional Orders."—*Local Government Chronicle.*

𝔚eale's 𝔚udimentary 𝔖eries.

LONDON, 1862.
THE PRIZE MEDAL
Was awarded to the Publishers of
"WEALE'S SERIES."

A NEW LIST OF
WEALE'S SERIES
RUDIMENTARY SCIENTIFIC, EDUCATIONAL, AND CLASSICAL.

Comprising nearly Three Hundred and Fifty distinct works in almost every department of Science, Art, and Education, recommended to the notice of Engineers, Architects, Builders, Artisans, and Students generally, as well as to those interested in Workmen's Libraries, Literary and Scientific Institutions, Colleges, Schools, Science Classes, &c., &c.

☞ " WEALE'S SERIES includes Text-Books on almost every branch of Science and Industry, comprising such subjects as Agriculture, Architecture and Building, Civil Engineering, Fine Arts, Mechanics and Mechanical Engineering, Physical and Chemical Science, and many miscellaneous Treatises. The whole are constantly undergoing revision, and new editions, brought up to the latest discoveries in scientific research, are constantly issued. The prices at which they are sold are as low as their excellence is assured."—*American Literary Gazette.*

" Amongst the literature of technical education, WEALE'S SERIES has ever enjoyed a high reputation, and the additions being made by Messrs. CROSBY LOCKWOOD & SON render the series more complete, and bring the information upon the several subjects down to the present time."—*Mining Journal.*

" It is not too much to say that no books have ever proved more popular with, or more useful to, young engineers and others than the excellent treatises comprised in WEALE'S SERIES."—*Engineer.*

" The excellence of WEALE'S SERIES is now so well appreciated, that it would be wasting our space to enlarge upon their general usefulness and value."—*Builder.*

" The volumes of WEALE'S SERIES form one of the best collections of elementary technical books in any language."—*Architect.*

" WEALE'S SERIES has become a standard as well as an unrivalled collection of treatises in all branches of art and science."—*Public Opinion.*

PHILADELPHIA, 1876.
THE PRIZE MEDAL
Was awarded to the Publishers for
Books : Rudimentary, Scientific,
"WEALE'S SERIES," ETC.

CROSBY LOCKWOOD & SON,
7, STATIONERS' HALL COURT, LUDGATE HILL, LONDON, E.C.

WEALE'S RUDIMENTARY SCIENTIFIC SERIES.

Capio Lumen

. The volumes of this Series are freely Illustrated with Woodcuts, or otherwise, where requisite. Throughout the following List it must be understood that the books are bound in limp cloth, unless otherwise stated; *but the volumes marked with a ‡ may also be had strongly bound in cloth boards for 6d. extra.*

N.B.—In ordering from this List it is recommended, as a means of facilitating business and obviating error, to quote the numbers affixed to the volumes, as well as the titles and prices.

CIVIL ENGINEERING, SURVEYING, ETC.

No.
31. *WELLS AND WELL-SINKING.* By JOHN GEO. SWINDELL, A.R.I.B.A., and G. R. BURNELL, C.E. Revised Edition. With a New Appendix on the Qualities of Water. Illustrated. 2s.
35. *THE BLASTING AND QUARRYING OF STONE,* for Building and other Purposes. By Gen. Sir J. BURGOYNE, Bart. 1s. 6d.
43. *TUBULAR, AND OTHER IRON GIRDER BRIDGES,* particularly describing the Britannia and Conway Tubular Bridges. By G. DRYSDALE DEMPSEY, C.E. Fourth Edition. 2s.
44. *FOUNDATIONS AND CONCRETE WORKS,* with Practical Remarks on Footings, Sand, Concrete, Béton, Pile-driving, Caissons, and Cofferdams, &c. By E. DOBSON. Fifth Edition. 1s. 6d.
60. *LAND AND ENGINEERING SURVEYING.* By T. BAKER, C.E. Fifteenth Edition, revised by Professor J. R. YOUNG. 2s.‡
80*. *EMBANKING LANDS FROM THE SEA.* With examples and Particulars of actual Embankments, &c. By J. WIGGINS, F.G.S. 2s.
81. *WATER WORKS,* for the Supply of Cities and Towns. With a Description of the Principal Geological Formations of England as influencing Supplies of Water, &c. By S. HUGHES, C.E. New Edition. 4s.‡
118. *CIVIL ENGINEERING IN NORTH AMERICA,* a Sketch of. By DAVID STEVENSON, F.R.S.E., &c. Plates and Diagrams. 3s.
167. *IRON BRIDGES, GIRDERS, ROOFS, AND OTHER WORKS.* By FRANCIS CAMPIN, C.E. 2s. 6d.‡
197. *ROADS AND STREETS.* By H. LAW, C.E., revised and enlarged by D. K. CLARK, C.E., including pavements of Stone, Wood, Asphalte, &c. 4s. 6d.‡
203. *SANITARY WORK IN THE SMALLER TOWNS AND IN VILLAGES.* By C. SLAGG, A.M.I.C.E. Revised Edition. 3s.‡
212. *GAS-WORKS, THEIR CONSTRUCTION AND ARRANGEMENT;* and the Manufacture and Distribution of Coal Gas. Originally written by SAMUEL HUGHES, C.E. Re-written and enlarged by WILLIAM RICHARDS, C.E. Seventh Edition, with important additions. 5s. 6d.‡
213. *PIONEER ENGINEERING.* A Treatise on the Engineering Operations connected with the Settlement of Waste Lands in New Countries. By EDWARD DOBSON, Assoc. Inst. C.E. 4s. 6d.‡
216. *MATERIALS AND CONSTRUCTION;* A Theoretical and Practical Treatise on the Strains, Designing, and Erection of Works of Construction. By FRANCIS CAMPIN, C.E. Second Edition, revised. 3s.‡
219. *CIVIL ENGINEERING.* By HENRY LAW, M.Inst. C.E. Including HYDRAULIC ENGINEERING by GEO. R. BURNELL, M.Inst. C.E. Seventh Edition, revised, with large additions by D. KINNEAR CLARK, M.Inst. C.E. 6s. 6d., Cloth boards, 7s. 6d.
268. *THE DRAINAGE OF LANDS, TOWNS, & BUILDINGS.* By G. D. DEMPSEY, C.E. Revised, with large Additions on Recent Practice in Drainage Engineering, by D. KINNEAR CLARK, M.I.C.E. Second Edition, Corrected. 4s. 6d.‡ [*Just published.*

☞ *The ‡ indicates that these vols. may be had strongly bound at 6d. extra.*

LONDON : CROSBY LOCKWOOD AND SON,

MECHANICAL ENGINEERING, ETC.

33. *CRANES, the Construction of, and other Machinery for Raising Heavy Bodies.* By Joseph Glynn, F.R.S. Illustrated. 1s. 6d.

34. *THE STEAM ENGINE.* By Dr. Lardner. Illustrated. 1s. 6d.

59. *STEAM BOILERS:* their Construction and Management. By R. Armstrong, C.E. Illustrated. 1s. 6d.

82. *THE POWER OF WATER,* as applied to drive Flour Mills, and to give motion to Turbines, &c. By Joseph Glynn, F.R.S. 2s.‡

98. *PRACTICAL MECHANISM,* the Elements of; and Machine Tools. By T. Baker, C.E. With Additions by J. Nasmyth, C.E. 2s. 6d.‡

139. *THE STEAM ENGINE,* a Treatise on the Mathematical Theory of, with Rules and Examples for Practical Men. By T. Baker, C.E. 1s. 6d

164. *MODERN WORKSHOP PRACTICE,* as applied to Steam Engines, Bridges, Ship-building, Cranes, &c. By J. G. Winton. Fourth Edition, much enlarged and carefully revised. 3s. 6d.‡ [*Just published.*

165. *IRON AND HEAT,* exhibiting the Principles concerned in the Construction of Iron Beams, Pillars, and Girders. By J. Armour. 2s. 6d.‡

166. *POWER IN MOTION:* Horse-Power, Toothed-Wheel Gearing, Long and Short Driving Bands, and Angular Forces. By J. Armour, 2s.‡

171. *THE WORKMAN'S MANUAL OF ENGINEERING DRAWING.* By J. Maxton. 6th Edn. With 7 Plates and 350 Cuts. 3s. 6d.‡

190. *STEAM AND THE STEAM ENGINE,* Stationary and Portable. Being an Extension of the Elementary Treatise on the Steam Engine of Mr. John Sewell. By D. K. Clark, M.I.C.E. 3s. 6d.‡

200. *FUEL,* its Combustion and Economy. By C. W. Williams. With Recent Practice in the Combustion and Economy of Fuel—Coal, Coke Wood, Peat, Petroleum, &c.—by D. K. Clark, M.I.C.E. 3s. 6d.‡

202. *LOCOMOTIVE ENGINES.* By G. D. Dempsey, C.E. ; with large additions by D. Kinnear Clark, M.I.C.E. 3s.‡

211. *THE BOILERMAKER'S ASSISTANT* in Drawing, Templating, and Calculating Boiler and Tank Work. By John Courtney. Practical Boiler Maker. Edited by D. K. Clark, C.E. 100 Illustrations. 2s,

217. *SEWING MACHINERY:* Its Construction, History, &c., with full Technical Directions for Adjusting, &c. By J. W. Urquhart, C.E. 2s.‡

223. *MECHANICAL ENGINEERING.* Comprising Metallurgy, Moulding, Casting, Forging. Tools, Workshop Machinery, Manufacture of the Steam Engine, &c. By Francis Campin, C.E. Second Edition. 2s. 6d.‡

236. *DETAILS OF MACHINERY.* Comprising Instructions for the Execution of various Works in Iron. By Francis Campin, C.E. 3s.‡

237. *THE SMITHY AND FORGE;* including the Farrier's Art and Coach Smithing. By W. J. E. Crane. Illustrated. 2s. 6d.‡

238. *THE SHEET-METAL WORKER'S GUIDE;* a Practical Handbook for Tinsmiths, Coppersmiths, Zincworkers, &c. With 94 Diagrams and Working Patterns. By W. J. E. Crane. Second Edition, revised. 1s. 5d.

251. *STEAM AND MACHINERY MANAGEMENT:* with Hints on Construction and Selection. By M. Powis Bale, M.I.M.E. 2s. 6d.‡

254. *THE BOILERMAKER'S READY-RECKONER.* By J. Courtney. Edited by D. K. Clark, C.E. 4s., limp; 5s., half-bound.

255. *LOCOMOTIVE ENGINE-DRIVING.* A Practical Manual for Engineers in charge of Locomotive Engines. By Michael Reynolds, M.S.E Eighth Edition. 3s. 6d., limp ; 4s. 6d. cloth boards.

256. *STATIONARY ENGINE-DRIVING.* A Practical Manual Engineers in charge of Stationary Engines. By Michael Reynolds, M.S.E. Third Edition. 3s. 6d. limp ; 4s. 6d. cloth boards.

260. *IRON BRIDGES OF MODERATE SPAN:* their Construction and Erection. By Hamilton W. Pendred, C.E. 2s.

☞ *The ‡ indicates that these vols. may be had strongly bound at 6d. extra.*

MINING, METALLURGY, ETC.

4. *MINERALOGY*, Rudiments of; a concise View of the General Properties of Minerals. By A. RAMSAY, F.G.S., F.R.G.S., &c. Third Edition, revised and enlarged. Illustrated. 3s. 6d.‡

117. *SUBTERRANEOUS SURVEYING*, with and without the Magnetic Needle. By T. FENWICK and T. BAKER, C.E. Illustrated. 2s. 6d. ‡

135. *ELECTRO-METALLURGY;* Practically Treated. By ALEXANDER WATT. Ninth Edition, enlarged and revised, with additional Illustrations, and including the most recent Processes. 3s. 6d.‡

172. *MINING TOOLS*, Manual of. For the Use of Mine Managers, Agents, Students, &c. By WILLIAM MORGANS. 2s. 6d.

172*. *MINING TOOLS, ATLAS* of Engravings to Illustrate the above, containing 235 Illustrations, drawn to Scale. 4to. 4s. 6d.

176. *METALLURGY OF IRON.* Containing History of Iron Manufacture, Methods of Assay, and Analyses of Iron Ores. Processes of Manufacture of Iron and Steel, &c. By H. BAUERMAN, F.G.S. Sixth Edition, revised and enlarged. 5s.‡ [*Just published.*

180. *COAL AND COAL MINING.* By the late Sir WARINGTON W. SMYTH, M.A., F.R.S. Seventh Edition, revised. 3s. 6d.‡ [*Just published.*

195. *THE MINERAL SURVEYOR AND VALUER'S COMPLETE GUIDE.* By W. LINTERN, M.E. Third Edition, including Magnetic and Angular Surveying. With Four Plates. 3s. 6d.‡

214. *SLATE AND SLATE QUARRYING*, Scientific, Practical, and Commercial. By D. C. DAVIES, F.G.S., Mining Engineer, &c. 3s.‡

264. *A FIRST BOOK OF MINING AND QUARRYING*, with the Sciences connected therewith, for Primary Schools and Self-Instruction. By J. H. COLLINS, F.G.S. Second Edition, with additions. 1s. 6d.

ARCHITECTURE, BUILDING, ETC.

16. *ARCHITECTURE—ORDERS*—The Orders and their Æsthetic Principles. By W. H. LEEDS. Illustrated. 1s. 6d.

17. *ARCHITECTURE—STYLES*—The History and Description of the Styles of Architecture of Various Countries, from the Earliest to the Present Period. By T. TALBOT BURY, F.R.I.B.A., &c. Illustrated. 2s.
  ⁎ ORDERS AND STYLES OF ARCHITECTURE, *in One Vol.*, 3s. 6d.

18. *ARCHITECTURE—DESIGN*—The Principles of Design in Architecture, as deducible from Nature and exemplified in the Works of the Greek and Gothic Architects. By E. L. GARBETT, Architect. Illustrated,.2s.6d.
•⁎• *The three preceding Works, in One handsome Vol., half bound, entitled* "MODERN ARCHITECTURE," *price* 6s.

22. *THE ART OF BUILDING*, Rudiments of. General Principles of Construction, Materials used in Building, Strength and Use of Materials, Working Drawings, Specifications, and Estimates. By E. DOBSON, 2s.‡

25. *MASONRY AND STONECUTTING:* Rudimentary Treatise on the Principles of Masonic Projection and their application to Construction. By EDWARD DOBSON, M.R.I.B.A., &c. 2s. 6d.‡

42. *COTTAGE BUILDING.* By C. BRUCE ALLEN, Architect. Tenth Edition, revised and enlarged. With a Chapter on Economic Cottages for Allotments, by EDWARD E. ALLEN, C.E. 2s.

45. *LIMES, CEMENTS, MORTARS, CONCRETES, MASTICS*, PLASTERING, &c. By G. R. BURNELL, C.E. Thirteenth Edition. 1s. 6d.

57. *WARMING AND VENTILATION.* An Exposition of the General Principles as applied to Domestic and Public Buildings, Mines, Lighthouses, Ships, &c. By C. TOMLINSON, F.R.S., &c. Illustrated. 3s.

111. *ARCHES, PIERS, BUTTRESSES, &c.:* Experimental Essays on the Principles of Construction. By W. BLAND. Illustrated. 1s. 6d.

☞ *The* ‡ *indicates that these vols. may be had strongly bound at 6d. extra.*

LONDON : CROSBY LOCKWOOD AND SON,

Architecture, Building, etc., *continued.*

116. *THE ACOUSTICS OF PUBLIC BUILDINGS;* or, The Principles of the Science of Sound applied to the purposes of the Architect and Builder. By T. Roger Smith, M.R.I.B.A., Architect. Illustrated. 1s. 6d.

127. *ARCHITECTURAL MODELLING IN PAPER,* the Art of. By T. A. Richardson, Architect. Illustrated. 1s. 6d.

128. *VITRUVIUS — THE ARCHITECTURE OF MARCUS VITRUVIUS POLLO.* In Ten Books. Translated from the Latin by Joseph Gwilt, F.S.A., F.R.A.S. With 23 Plates. 5s.

130. *GRECIAN ARCHITECTURE,* An Inquiry into the Principles of Beauty in; with an Historical View of the Rise and Progress of the Art in Greece. By the Earl of Aberdeen. 1s.

*** *The two preceding Works in One handsome Vol., half bound, entitled "*Ancient Architecture," *price 6s.*

132. *THE ERECTION OF DWELLING-HOUSES.* Illustrated by a Perspective View, Plans, Elevations, and Sections of a pair of Semi-detached Villas, with the Specification, Quantities, and Estimates, &c. By S. H. Brooks. New Edition. with Plates. 2s. 6d.‡

156. *QUANTITIES & MEASUREMENTS* in Bricklayers', Masons', Plasterers', Plumbers', Painters', Paperhangers', Gilders', Smiths', Carpenters' and Joiners' Work. By A. C. Heaton, Surveyor. New Edition. 1s. 6d.

175. *LOCKWOOD'S BUILDER'S PRICE BOOK FOR* 1891. A Comprehensive Handbook of the Latest Prices and Data for Builders, Architects, Engineers, and Contractors. Re-constructed, Re-written, and greatly Enlarged. By Francis T. W. Miller, A.R.I.B.A. 650 pages. 3s. 6d. ; cloth hoards, 4s. [*Just Published.*

182. *CARPENTRY AND JOINERY*—The Elementary Principles of Carpentry. Chiefly composed from the Standard Work of Thomas Tredgold, C.E. With a Treatise on Joinery by E. Wyndham Tarn, M.A. Fifth Edition, Revised. 3s. 6d.‡

182*. *CARPENTRY AND JOINERY. ATLAS* of 35 Plates to accompany the above. With Descriptive Letterpress. 4to. 6s.

185. *THE COMPLETE MEASURER ;* the Measurement of Boards, Glass, &c. ; Unequal-sided, Square-sided, Octagonal-sided, Round Timber and Stone, and Standing Timber, &c. By Richard Horton. Fifth Edition. 4s. ; strongly bound in leather, 5s.

187. *HINTS TO YOUNG ARCHITECTS.* By G. Wightwick. New Edition. By G. H. Guillaume. Illustrated. 3s. 6d.‡

188. *HOUSE PAINTING, GRAINING, MARBLING, AND SIGN WRITING :* with a Course of Elementary Drawing for House-Painters, Sign-Writers, &c., and a Collection of Useful Receipts. By Ellis A. Davidson. Sixth Edition. With Coloured Plates 5s. cloth limp ; 6s. cloth hoards.

189. *THE RUDIMENTS OF PRACTICAL BRICKLAYING.* In Six Sections : General Principles ; Arch Drawing, Cutting, and Setting ; Pointing ; Paving, Tiling, Materials ; Slating and Plastering ; Practical Geometry, Mensuration, &c. By Adam Hammond. Seventh Edition. 1s. 6d.

191. *PLUMBING.* A Text-Book to the Practice of the Art or Craft of the Plumber. With Chapters upon House Drainage and Ventilation. Fifth Edition. With 380 Illustrations. By W. P. Buchan. 3s. 6d.‡

192. *THE TIMBER IMPORTER'S, TIMBER MERCHANT'S,* and BUILDER'S STANDARD GUIDE. By R. E. Grandy. 2s.

206. *A BOOK ON BUILDING, Civil and Ecclesiastical,* including Church Restoration. With the Theory of Domes and the Great Pyramid, &c. By Sir Edmund Beckett, Bart., LL.D., Q.C., F.R.A.S. 4s. 6d.‡

226. *THE JOINTS MADE AND USED BY BUILDERS* in the Construction of various kinds of Engineering and Architectural Works. By Wyvill J. Christy, Architect. With upwards of 160 Engravings on Wood. 3s.‡

228. *THE CONSTRUCTION OF ROOFS OF WOOD AND IRON.* By E. Wyndham Tarn, M.A., Architect. Second Edition, revised. 1s. 6d.

☞ *The ‡ indicates that these vols. may be had strongly bound at 6d. extra.*

Architecture, Building, etc., *continued.*

229. *ELEMENTARY DECORATION:* as applied to the Interior and Exterior Decoration of Dwelling-Houses, &c. By J. W. FACEY. 2s.

257. *PRACTICAL HOUSE DECORATION.* A Guide to the Art of Ornamental Painting. By JAMES W. FACEY. 2s. 6d.
. *The two preceding Works, in One handsome Vol., half-bound, entitled "* HOUSE DECORATION, ELEMENTARY AND PRACTICAL," *price* 5s.

230. *HANDRAILING.* Showing New and Simple Methods for finding the Pitch of the Plank. Drawing the Moulds, Bevelling, Jointing-up, and Squaring the Wreath. By GEORGE COLLINGS. Second Edition, Revised including A TREATISE ON STAIRBUILDING. Plates and Diagrams. 2s. 6d.

247. *BUILDING ESTATES:* a Rudimentary Treatise on the Development, Sale, Purchase, and General Management of Building Land. By FOWLER MAITLAND, Surveyor. Second Edition, revised. 2s.

248. *PORTLAND CEMENT FOR USERS.* By HENRY FAIJA, Assoc. M. Inst. C.E. Third Edition, corrected. Illustrated. 2s.

252. *BRICKWORK:* a Practical Treatise, embodying the General and Higher Principles of Bricklaying, Cutting and Setting, &c. By F. WALKER. Second Edition, Revised and Enlarged. 1s. 6d.

23. *THE PRACTICAL BRICK AND TILE BOOK.* Comprising :
189. BRICK AND TILE MAKING, by E. DOBSON, A.I.C.E.; PRACTICAL BRICKLAY-
265. ING, by A. HAMMOND; BRICKCUTTING AND SETTING, by A. HAMMOND. 534 pp. with 270 Illustrations. 6s. Strongly half-bound.

253. *THE TIMBER MERCHANT'S, SAW-MILLER'S, AND* IMPORTER'S FREIGHT-BOOK AND ASSISTANT. By WM. RICH-ARDSON. With a Chapter on Speeds of Saw-Mill Machinery, &c. By M. POWIS BALE, A.M.Inst.C.E. 3s.‡

258. *CIRCULAR WORK IN CARPENTRY AND JOINERY.* A Practical Treatise on Circular Work of Single and Double Curvature. By GEORGE COLLINGS. Second Edition, 2s. 6d.

259. *GAS FITTING:* A Practical Handbook treating of every Description of Gas Laying and Fitting. By JOHN BLACK. With 122 Illustrations. 2s. 6d.‡

261. *SHORING AND ITS APPLICATION:* A Handbook for the Use of Students. By GEORGE H. BLAGROVE. 1s. 6d. [*Just published.*

265. *THE ART OF PRACTICAL BRICK CUTTING & SETTING.* By ADAM HAMMOND. With 90 Engravings. 1s. 6d. [*Just published.*

267. *THE SCIENCE OF BUILDING:* An Elementary Treatise on the Principles of Construction. Adapted to the Requirements of Architectural Students. By E. WYNDHAM TARN, M.A. Lond. Third Edition, Revised and Enlarged. With 59 Wood Engravings. 3s. 6d.‡ [*Just published.*

271. *VENTILATION:* a Text-book to the Practice of the Art of Ventilating Buildings, with a Supplementary Chapter upon Air Testing. By WILLIAM PATON BUCHAN, R.P., Sanitary and Ventilating Engineer, Author of "Plumbing," &c. 3s. 6d.‡ [*Just published.*

SHIPBUILDING, NAVIGATION, MARINE ENGINEERING, ETC.

51. *NAVAL ARCHITECTURE.* An Exposition of the Elementary Principles of the Science, and their Practical Application to Naval Construction. By J. PEAKE. Fifth Edition, with Plates and Diagrams. 3s. 6d.‡

53*. *SHIPS FOR OCEAN & RIVER SERVICE,* Elementary and Practical Principles of the Construction of. By H. A. SOMMERFELDT. 1s. 6d.

53**. *AN ATLAS OF ENGRAVINGS* to Illustrate the above. Twelve large folding plates. Royal 4to, cloth. 7s. 6d.

54. *MASTING, MAST-MAKING, AND RIGGING OF SHIPS,* Also Tables of Spars, Rigging, Blocks; Chain, Wire, and Hemp Ropes, &c., relative to every class of vessels. By ROBERT KIPPING, N.A. 2s.

☞ *The ‡ indicates that these vols. may be had strongly bound at 6d. extra.*

Shipbuilding, Navigation, Marine Engineering, etc., *cont.*

54*. *IRON SHIP-BUILDING.* With Practical Examples and Details.
By John Grantham, C.E. Fifth Edition. 4s.

55. *THE SAILOR'S SEA BOOK:* a Rudimentary Treatise on
Navigation. By James Greenwood, B.A. With numerous Woodcuts and
Coloured Plates. New and enlarged edition. By W. H. Rosser. 2s. 6d.‡

80. *MARINE ENGINES AND STEAM VESSELS.* By Robert
Murray, C.E. Eighth Edition, thoroughly Revised, with Additions by the
Author and by George Carlisle, C.E. 4s. 6d. limp; 5s. cloth boards.

83bis. *THE FORMS OF SHIPS AND BOATS.* By W. Bland.
Seventh Edition, Revised, with numerous Illustrations and Models. 1s. 6d.

99. *NAVIGATION AND NAUTICAL ASTRONOMY*, in Theory
and Practice. By Prof. J. R. Young. New Edition. 2s. 6d.

106. *SHIPS' ANCHORS*, a Treatise on. By G. Cotsell, N.A. 1s. 6d.

149. *SAILS AND SAIL-MAKING.* With Draughting, and the Centre
of Effort of the Sails; Weights and Sizes of Ropes: Masting, Rigging,
and Sails of Steam Vessels, &c. 12th Edition. By R. Kipping, N.A., 2s 6d.‡

155. *ENGINEER'S GUIDE TO THE ROYAL & MERCANTILE*
NAVIES. By a Practical Engineer. Revised by D. F. M'Carthy. 3s.

55
& *PRACTICAL NAVIGATION.* Consisting of The Sailor's
204. Sea-Book. By James Greenwood and W. H. Rosser. Together with
the requisite Mathematical and Nautical Tables for the Working of the
Problems. By H. Law, C.E., and Prof. J. R. Young. 7s. Half-bound.

AGRICULTURE, GARDENING, ETC.

61*. *A COMPLETE READY RECKONER FOR THE ADMEA-*
SUREMENT OF LAND, &c. By A. Arman. Third Edition, revised
and extended by C. Norris, Surveyor, Valuer, &c. 2s.

131. *MILLER'S, CORN MERCHANT'S, AND FARMER'S*
READY RECKONER. Second Edition, with a Price List of Modern
Flour-Mill Machinery, by W. S. Hutton, C.E. 2s.

140. *SOILS, MANURES, AND CROPS.* (Vol. 1. Outlines of
Modern Farming.) By R. Scott Burn. Woodcuts. 2s.

141. *FARMING & FARMING ECONOMY*, Notes, Historical and
Practical, on. (Vol. 2. Outlines of Modern Farming.) By R. Scott Burn. 3s.

142. *STOCK; CATTLE, SHEEP, AND HORSES.* (Vol. 3.
Outlines of Modern Farming.) By R. Scott Burn. Woodcuts. 2s. 6d.

145. *DAIRY, PIGS, AND POULTRY*, Management of the. By
R. Scott Burn. (Vol. 4. Outlines of Modern Farming.) 2s.

146. *UTILIZATION OF SEWAGE, IRRIGATION, AND*
RECLAMATION OF WASTE LAND. (Vol. 5. Outlines of Modern
Farming.) By R. Scott Burn. Woodcuts. 2s. 6d.

⁎ Nos. 140-1-2-5-6, *in One Vol., handsomely half-bound, entitled "*Outlines of
Modern Farming.*" By Robert Scott Burn. Price* 12s.

177. *FRUIT TREES*, The Scientific and Profitable Culture of. From
the French of Du Breuil. Revised by Geo. Glenny. 187 Woodcuts. 3s. 6d.‡

198. *SHEEP; THE HISTORY, STRUCTURE, ECONOMY, AND*
DISEASES OF. By W. C. Spooner, M.R.V.C., &c. Fifth Edition,
enlarged, including Specimens of New and Improved Breeds. 3s. 6d.‡

201. *KITCHEN GARDENING MADE EASY.* By George M. F.
Glenny. Illustrated. 1s. 6d.‡

207. *OUTLINES OF FARM MANAGEMENT, and the Organi-*
zation of Farm Labour. By R. Scott Burn. 2s. 6d.‡

208. *OUTLINES OF LANDED ESTATES MANAGEMENT*
By R. Scott Burn. 2s. 6d.

⁎ Nos. 207 & 208 *in One Vol., handsomely half-bound, entitled "*Outlines of
Landed Estates and Farm Management.*" By R. Scott Burn. Price* 6s.

☞ *The ‡ indicates that these vols. may be had strongly bound at 6d. extra.*

Agriculture, Gardening, etc., *continued.*

209. *THE TREE PLANTER AND PLANT PROPAGATOR.*
A Practical Manual on the Propagation of Forest Trees, Fruit Trees, Flowering Shrubs, Flowering Plants, &c. By SAMUEL WOOD. 2s.

210. *THE TREE PRUNER.* A Practical Manual on the Pruning of Fruit Trees, including also their Training and Renovation; also the Pruning of Shrubs, Climbers, and Flowering Plants. By SAMUEL WOOD. 1s. 6d.

** *Nos.* 209 & 210 *in One Vol., handsomely half-bound, entitled* "THE TREE PLANTER, PROPAGATOR, AND PRUNER." By SAMUEL WOOD. *Price* 3s. 6d.

218. *THE HAY AND STRAW MEASURER:* Being New Tables for the Use of Auctioneers, Valuers, Farmers, Hay and Straw Dealers, &c. By JOHN STEELE. Fourth Edition. 2s.

222. *SUBURBAN FARMING.* The Laying-out and Cultivation of Farms, adapted to the Produce of Milk, Butter, and Cheese, Eggs, Poultry, and Pigs. By Prof. JOHN DONALDSON and R. SCOTT BURN. 3s. 6d.‡

231. *THE ART OF GRAFTING AND BUDDING.* By CHARLES BALTET. With Illustrations. 2s. 6d.‡

232. *COTTAGE GARDENING;* or, Flowers, Fruits, and Vegetables for Small Gardens. By E. HOBDAY. 1s. 6d.

233. *GARDEN RECEIPTS.* Edited by CHARLES W. QUIN. 1s. 6d.

234. *MARKET AND KITCHEN GARDENING.* By C. W. SHAW, late Editor of "Gardening Illustrated." 3s.‡   [*Just published.*

239. *DRAINING AND EMBANKING.* A Practical Treatise, embodying the most recent experience in the Application of Improved Methods. By JOHN SCOTT, late Professor of Agriculture and Rural Economy at the Royal Agricultural College, Cirencester. With 68 Illustrations. 1s. 6d.

240. *IRRIGATION AND WATER SUPPLY.* A Treatise on Water Meadows, Sewage Irrigation, and Warping; the Construction of Wells, Ponds, and Reservoirs, &c. By Prof. JOHN SCOTT. With 34 Illus. 1s. 6d.

241. *FARM ROADS, FENCES, AND GATES.* A Practical Treatise on the Roads, Tramways, and Waterways of the Farm; the Principles of Enclosures; and the different kinds of Fences, Gates, and Stiles. By Professor JOHN SCOTT. With 75 Illustrations. 1s. 6d.

242. *FARM BUILDINGS.* A Practical Treatise on the Buildings necessary for various kinds of Farms, their Arrangement and Construction, with Plans and Estimates. By Prof. JOHN SCOTT. With 105 Illus. 2s.

243. *BARN IMPLEMENTS AND MACHINES.* A Practical Treatise on the Application of Power to the Operations of Agriculture; and on various Machines used in the Threshing-barn, in the Stock-yard, and in the Dairy, &c. By Prof. J. SCOTT. With 123 Illustrations. 2s.

244. *FIELD IMPLEMENTS AND MACHINES.* A Practical Treatise on the Varieties now in use, with Principles and Details of Construction, their Points of Excellence, and Management. By Professor JOHN SCOTT. With 138 Illustrations. 2s.

245. *AGRICULTURAL SURVEYING.* A Practical Treatise on Land Surveying, Levelling, and Setting-out; and on Measuring and Estimating Quantities, Weights, and Values of Materials, Produce, Stock, &c. By Prof. JOHN SCOTT. With 62 Illustrations. 1s. 6d.

** *Nos.* 239 *to* 245 *in One Vol., handsomely half-bound, entitled* "THE COMPLETE TEXT-BOOK OF FARM ENGINEERING." By Professor JOHN SCOTT. *Price* 12s.

250. *MEAT PRODUCTION.* A Manual for Producers, Distributors, &c. By JOHN EWART. 2s. 6d.‡

266. *BOOK-KEEPING FOR FARMERS & ESTATE OWNERS.* By J. M. WOODMAN, Chartered Accountant. 2s. 6d. cloth limp; 3s. 6d. cloth boards.   [*Just published.*

☞ *The* ‡ *indicates that these vols. may be had strongly bound at 6d. extra.*

LONDON : CROSBY LOCKWOOD AND SON.

MATHEMATICS, ARITHMETIC, ETC.

32. *MATHEMATICAL INSTRUMENTS*, a Treatise on; Their Construction, Adjustment, Testing, and Use concisely Explained. By J. F. HEATHER, M.A. Fourteenth Edition, revised, with additions, by A. T. WALMISLEY, M.I.C.E., Fellow of the Surveyors' Institution. Original Edition, in 1 vol., Illustrated. 2s.‡ [*Just published.*

* *In ordering the above, be careful to say, " Original Edition " (No. 32), to distinguish it from the Enlarged Edition in 3 vols. (Nos. 168-9-70.)*

76. *DESCRIPTIVE GEOMETRY*, an Elementary Treatise on ; with a Theory of Shadows and of Perspective, extracted from the French of G. MONGE. To which is added, a description of the Principles and Practice of Isometrical Projection. By J. F. HEATHER, M.A. With 14 Plates. 2s.

178. *PRACTICAL PLANE GEOMETRY:* giving the Simplest Modes of Constructing Figures contained in one Plane and Geometrical Construction of the Ground. By J. F. HEATHER, M.A. With 215 Woodcuts. 2s.

83. *COMMERCIAL BOOK-KEEPING.* With Commercial Phrases and Forms in English, French, Italian, and German. By JAMES HADDON, M.A., Arithmetical Master of King's College School, London. 1s. 6d.

84. *ARITHMETIC*, a Rudimentary Treatise on : with full Explanations of its Theoretical Principles, and numerous Examples for Practice. By Professor J. R. YOUNG. Eleventh Edition. 1s. 6d.

84*. A KEY to the above, containing Solutions in full to the Exercises, together with Comments, Explanations, and Improved Processes, for the Use of Teachers and Unassisted Learners. By J. R. YOUNG. 1s. 6d.

85. *EQUATIONAL ARITHMETIC*, applied to Questions of Interest, Annuities, Life Assurance, and General Commerce ; with various Tables by which all Calculations may be greatly facilitated. By W. HIPSLEY. 2s.

86. *ALGEBRA*, the Elements of. By JAMES HADDON, M.A. With Appendix, containing miscellaneous Investigations, and a Collection of Problems in various parts of Algebra. 2s.

86*. A KEY AND COMPANION to the above Book, forming an extensive repository of Solved Examples and Problems in Illustration of the various Expedients necessary in Algebraical Operations. By J. R. YOUNG. 1s. 6d.

88. *EUCLID*, THE ELEMENTS OF : with many additional Propositions
89. and Explanatory Notes : to which is prefixed, an Introductory Essay on Logic. By HENRY LAW, C.E. 2s. 6d.‡

* *Sold also separately, viz. :—*

88. EUCLID, The First Three Books. By HENRY LAW, C.E. 1s. 6d.
89. EUCLID, Books 4, 5, 6, 11, 12. By HENRY LAW, C.E. 1s. 6d.

90. *ANALYTICAL GEOMETRY AND CONIC SECTIONS*, By JAMES HANN. A New Edition, by Professor J. R. YOUNG. 2s.‡

91. *PLANE TRIGONOMETRY*, the Elements of. By JAMES HANN, formerly Mathematical Master of King's College, London. 1s. 6d.

92. *SPHERICAL TRIGONOMETRY*, the Elements of. By JAMES HANN. Revised by CHARLES H. DOWLING, C.E. 1s.
* *Or with " The Elements of Plane Trigonometry," in One Volume, 2s. 6d.*

93. *MENSURATION AND MEASURING.* With the Mensuration and Levelling of Land for the Purposes of Modern Engineering. By T. BAKER, C.E. New Edition by E. NUGENT, C.E. Illustrated. 1s. 6d.

101. *DIFFERENTIAL CALCULUS*, Elements of the. By W. S. B. WOOLHOUSE, F.R.A.S., &c. 1s. 6d.

102. *INTEGRAL CALCULUS*, Rudimentary Treatise on the. By HOMERSHAM COX, B.A. Illustrated. 1s.

136. *ARITHMETIC*, Rudimentary, for the Use of Schools and Self-Instruction. By JAMES HADDON, M.A. Revised by A. ARMAN. 1s. 6d.
137. A KEY TO HADDON'S RUDIMENTARY ARITHMETIC. By A. ARMAN. 1s. 6d.

☞ *The ‡ indicates that these vols. may be had strongly bound at 6d. extra.*

Mathematics, Arithmetic, etc., *continued.*

168. *DRAWING AND MEASURING INSTRUMENTS.* Including—I. Instruments employed in Geometrical and Mechanical Drawing, and in the Construction, Copying, and Measurement of Maps and Plans. II. Instruments used for the purposes of Accurate Measurement, and for Arithmetical Computations. By J. F. HEATHER, M.A. Illustrated. 1s. 6d

169. *OPTICAL INSTRUMENTS.* Including (more especially) Telescopes, Microscopes, and Apparatus for producing copies of Maps and Plans by Photography. By J. F. HEATHER, M.A. Illustrated. 1s. 6d.

170. *SURVEYING AND ASTRONOMICAL INSTRUMENTS.* Including—I. Instruments Used for Determining the Geometrical Features of a portion of Ground. II. Instruments Employed in Astronomical Observations. By J. F. HEATHER, M.A. Illustrated. 1s. 6d.

✱ *The above three volumes form an enlargement of the Author's original work "Mathematical Instruments." (See No. 32 in the Series.)*

168.⎫ *MATHEMATICAL INSTRUMENTS.* By J. F. HEATHER,
169.⎬ M.A. Enlarged Edition, for the most part entirely re-written. The 3 Parts as
170.⎭ above, in One thick Volume. With numerous Illustrations. 4s. 6d.‡

158. *THE SLIDE RULE, AND HOW TO USE IT;* containing full, easy, and simple Instructions to perform all Business Calculations with unexampled rapidity and accuracy. By CHARLES HOARE, C.E. Fifth Edition. With a Slide Rule in tuck of cover. 2s. 6d.‡

196. *THEORY OF COMPOUND INTEREST AND ANNUITIES;* with Tables of Logarithms for the more Difficult Computations of Interest, Discount, Annuities, &c. By FÉDOR THOMAN. 4s.‡

199. *THE COMPENDIOUS CALCULATOR ;* or, Easy and Concise Methods of Performing the various Arithmetical Operations required in Commercial and Business Transactions ; together with Useful Tables. By D. O'GORMAN. Twenty-seventh Edition, carefully revised by C. NORRIS. 2s. 6d., cloth limp; 3s. 6d., strongly half-bound in leather.

204. *MATHEMATICAL TABLES,* for Trigonometrical, Astronomical, and Nautical Calculations ; to which is prefixed a Treatise on Logarithms. By HENRY LAW, C.E. Together with a Series of Tables for Navigation and Nautical Astronomy. By Prof. J. R. YOUNG. New Edition. 4s.

204✱. *LOGARITHMS.* With Mathematical Tables for Trigonometrical, Astronomical, and Nautical Calculations. By HENRY LAW, M.Inst.C.E. New and Revised Edition. (Forming part of the above Work). 3s.

221. *MEASURES, WEIGHTS, AND MONEYS OF ALL NATIONS,* and an Analysis of the Christian, Hebrew, and Mahometan Calendars. By W. S. B. WOOLHOUSE, F.R.A.S., F.S.S. Seventh Edition, 2s. 6d.‡

227. *MATHEMATICS AS APPLIED TO THE CONSTRUCTIVE ARTS.* Illustrating the various processes of Mathematical Investigation, by means of Arithmetical and Simple Algebraical Equations and Practical Examples. By FRANCIS CAMPIN, C.E. Second Edition. 3s.‡

PHYSICAL SCIENCE, NATURAL PHILOSOPHY, ETC.

1. *CHEMISTRY.* By Professor GEORGE FOWNES, F.R.S. With an Appendix on the Application of Chemistry to Agriculture. 1s.

2. *NATURAL PHILOSOPHY,* Introduction to the Study of. By C. TOMLINSON. Woodcuts. 1s. 6d.

6. *MECHANICS,* Rudimentary Treatise on. By CHARLES TOMLINSON. Illustrated. 1s. 6d.

7. *ELECTRICITY;* showing the General Principles of Electrical Science, and the purposes to which it has been applied. By Sir W. SNOW HARRIS, F.R.S., &c. With Additions by R. SABINE, C.E., F.S.A. 1s. 6d.

7✱. *GALVANISM.* By Sir W. SNOW HARRIS. New Edition by ROBERT SABINE, C.E., F.S.A. 1s. 6d.

8. *MAGNETISM;* being a concise Exposition of the General Principles of Magnetical Science. By Sir W. SNOW HARRIS. New Edition, revised by H. M. NOAD, Ph.D. With 165 Woodcuts. 3s. 6d.‡

☞ *The ‡ indicates that these vols. may be had strongly bound at 6d. extra.*

Physical Science, Natural Philosophy, etc., *continued.*

11. *THE ELECTRIC TELEGRAPH;* its History and Progress; with Descriptions of some of the Apparatus. By R. SABINE, C.E., F.S.A. 3s.

12. *PNEUMATICS,* including Acoustics and the Phenomena of Wind Currents, for the Use of Beginners By CHARLES TOMLINSON, F.R.S. Fourth Edition, enlarged. Illustrated. 1s. 6d. [*Just published.*

72. *MANUAL OF THE MOLLUSCA;* a Treatise on Recent and Fossil Shells. By Dr. S. P. WOODWARD, A.L.S. Fourth Edition. With Plates and 300 Woodcuts. 7s. 6d., cloth.

96. *ASTRONOMY.* By the late Rev. ROBERT MAIN, M.A. Third Edition, by WILLIAM THYNNE LYNN, B.A., F.R.A.S. 2s.

97. *STATICS AND DYNAMICS,* the Principles and Practice of; embracing also a clear development of Hydrostatics, Hydrodynamics, and Central Forces. By T. BAKER, C.E. Fourth Edition. 1s. 6d.

173. *PHYSICAL GEOLOGY,* partly based on Major-General PORT-LOCK's "Rudiments of Geology." By RALPH TATE, A.L.S., &c. Woodcuts. 2s.

174. *HISTORICAL GEOLOGY,* partly based on Major-General PORTLOCK's "Rudiments." By RALPH TATE, A.L.S., &c. Woodcuts. 2s. 6d.

173 & 174. *RUDIMENTARY TREATISE ON GEOLOGY,* Physical and Historical. Partly based on Major-General PORTLOCK's "Rudiments of Geology." By RALPH TATE, A.L.S., F.G.S., &c. In One Volume. 4s. 6d.‡

183 & 184. *ANIMAL PHYSICS,* Handbook of. By Dr. LARDNER, D.C.L., formerly Professor of Natural Philosophy and Astronomy in University College, Lond. With 520 Illustrations. In One Vol. 7s. 6d., cloth boards.
  * *Sold also in Two Parts, as follows :—*

183. ANIMAL PHYSICS. By Dr. LARDNER. Part I., Chapters I.—VII. 4s.

184. ANIMAL PHYSICS. By Dr. LARDNER. Part II., Chapters VIII.—XVIII. 3s.

269. *LIGHT:* an Introduction to the Science of Optics, for the Use of Students of Architecture, Engineering, and other Applied Sciences. By E. WYNDHAM TARN, M.A. 1s. 6d. [*Just published.*

FINE ARTS.

20. *PERSPECTIVE FOR BEGINNERS.* Adapted to Young Students and Amateurs in Architecture, Painting, &c. By GEORGE PYNE. 2s.

40 *GLASS STAINING, AND THE ART OF PAINTING ON GLASS.* From the German of Dr. GESSERT and EMANUEL OTTO FROMBERG. With an Appendix on THE ART OF ENAMELLING. 2s. 6d.

69. *MUSIC,* A Rudimentary and Practical Treatise on. With numerous Examples. By CHARLES CHILD SPENCER. 2s. 6d.

71. *PIANOFORTE,* The Art of Playing the. With numerous Exercises & Lessons from the Best Masters. By CHARLES CHILD SPENCER. 1s. 6d.

69-71. *MUSIC & THE PIANOFORTE.* In one vol. Half bound, 5s.

181. *PAINTING POPULARLY EXPLAINED,* including Fresco, Oil, Mosaic, Water Colour, Water-Glass, Tempera, Encaustic, Miniature, Painting on Ivory, Vellum, Pottery, Enamel, Glass, &c. With Historical Sketches of the Progress of the Art by THOMAS JOHN GULLICK, assisted by JOHN TIMBS, F.S.A. Fifth Edition, revised and enlarged. 5s.‡

186. *A GRAMMAR OF COLOURING,* applied to Decorative Painting and the Arts. By GEORGE FIELD. New Edition, enlarged and adapted to the Use of the Ornamental Painter and Designer. By ELLIS A. DAVIDSON. With two new Coloured Diagrams, &c. 3s.‡

246. *A DICTIONARY OF PAINTERS, AND HANDBOOK FOR PICTURE AMATEURS;* including Methods of Painting, Cleaning, Re-lining and Restoring, Schools of Painting, &c. With Notes on the Copyists and Imitators of each Master. By PHILIPPE DARYL. 2s. 6d.‡

The ‡ indicates that these vols. may be had strongly bound at 6d. extra.

INDUSTRIAL AND USEFUL ARTS.

23. *BRICKS AND TILES*, Rudimentary Treatise on the Manufacture of. By E. DOBSON, M.R.I.B.A. Illustrated, 3s.‡
67. *CLOCKS, WATCHES, AND BELLS*, a Rudimentary Treatise on. By Sir EDMUND BECKETT, LL.D., Q.C. Seventh Edition, revised and enlarged. 4s. 6d. limp; 5s. 6d. cloth boards.
83**. *CONSTRUCTION OF DOOR LOCKS*. Compiled from the Papers of A. C. HOBBS, and Edited by CHARLES TOMLINSON. F.R.S. 2s. 6d.
162. *THE BRASS FOUNDER'S MANUAL;* Instructions for Modelling, Pattern-Making, Moulding, Turning, Filing, Burnishing, Bronzing, &c. With copious Receipts. &c. By WALTER GRAHAM. 2s.‡
205. *THE ART OF LETTER PAINTING MADE EASY.* By J. G. BADENOCH. Illustrated with 12 full-page Engravings of Examples. 1s. 6d.
215. *THE GOLDSMITH'S HANDBOOK*, containing full Instructions for the Alloying and Working of Gold. By GEORGE E. GEE, 3s.‡
225. *THE SILVERSMITH'S HANDBOOK*, containing full Instructions for the Alloying and Working of Silver. By GEORGE E. GEE. 3s.‡
. *The two preceding Works, in One handsome Vol., half-bound, entitled "*THE GOLDSMITH'S & SILVERSMITH'S COMPLETE HANDBOOK," 7s.
249. *THE HALL-MARKING OF JEWELLERY PRACTICALLY CONSIDERED.* By GEORGE E. GEE. 3s.‡
224. *COACH BUILDING*, A Practical Treatise, Historical and Descriptive. By J. W. BURGESS. 2s. 6d.‡
235. *PRACTICAL ORGAN BUILDING.* By W. E. DICKSON, M.A., Precentor of Ely Cathedral. Illustrated. 2s. 6d.‡
262. *THE ART OF BOOT AND SHOEMAKING.* By JOHN BEDFORD LENO. Numerous Illustrations. Third Edition. 2s.
263. *MECHANICAL DENTISTRY:* A Practical Treatise on the Construction of the Various Kinds of Artificial Dentures, with Formulæ, Tables, Receipts, &c. By CHARLES HUNTER. Third Edition. 3s.‡
270. *WOOD ENGRAVING:* A Practical and Easy Introduction to the Study of the Art. By W. N. BROWN. 1s. 6d.

MISCELLANEOUS VOLUMES.

36. *A DICTIONARY OF TERMS used in ARCHITECTURE, BUILDING, ENGINEERING, MINING, METALLURGY, ARCHÆOLOGY, the FINE ARTS, &c.* By JOHN WEALE. Fifth Edition. Revised by ROBERT HUNT, F.R.S. Illustrated. 5s. limp; 6s. cloth boards.
50. *THE LAW OF CONTRACTS FOR WORKS AND SERVICES.* By DAVID GIBBONS. Third Edition, enlarged. 3s.‡
112. *MANUAL OF DOMESTIC MEDICINE.* By R. GOODING, B.A., M.D. A Family Guide in all Cases of Accident and Emergency 2s.
112*. *MANAGEMENT OF HEALTH.* A Manual of Home and Personal Hygiene. By the Rev. JAMES BAIRD, B.A. 1s.
150. *LOGIC*, Pure and Applied. By S. H. EMMENS. 1s. 6d.
153. *SELECTIONS FROM LOCKE'S ESSAYS ON THE HUMAN UNDERSTANDING.* With Notes by S. H. EMMENS. 2s.
154. *GENERAL HINTS TO EMIGRANTS.* 2s.
157. *THE EMIGRANT'S GUIDE TO NATAL.* By ROBERT JAMES MANN, F.R.A.S., F.M.S. Second Edition. Map. 2s.
193. *HANDBOOK OF FIELD FORTIFICATION.* By Major W. W. KNOLLYS, F.R.G.S. With 163 Woodcuts. 3s.‡
194. *THE HOUSE MANAGER:* Being a Guide to Housekeeping. Practical Cookery, Pickling and Preserving, Household Work, Dairy Management, &c. By AN OLD HOUSEKEEPER. 3s. 6d.‡
194, *HOUSE BOOK (The).* Comprising :—I. THE HOUSE MANAGER.
112 & By an OLD HOUSEKEEPER. II. DOMESTIC MEDICINE. By R. GOODING, M.D.
112*. III. MANAGEMENT OF HEALTH. By J. BAIRD. In One Vol., half-bound, 6s.

☞ *The ‡ indicates that these vols may be had strongly bound at 6d. extra.*

LONDON : CROSBY LOCKWOOD AND SON.

EDUCATIONAL AND CLASSICAL SERIES.

HISTORY.

1. **England, Outlines of the History of;** more especially with reference to the Origin and Progress of the English Constitution. By WILLIAM DOUGLAS HAMILTON, F.S.A., of Her Majesty's Public Record Office. 4th Edition, revised. 5s.; cloth boards, 6s.

5. **Greece, Outlines of the History of;** in connection with the Rise of the Arts and Civilization in Europe. By W. DOUGLAS HAMILTON, of University College, London, and EDWARD LEVIEN, M.A., of Balliol College, Oxford. 2s. 6d.; cloth boards, 3s. 6d.

7. **Rome, Outlines of the History of:** from the Earliest Period to the Christian Era and the Commencement of the Decline of the Empire. By EDWARD LEVIEN, of Balliol College, Oxford. Map, 2s. 6d.; cl. hds. 3s. 6d.

9. **Chronology of History, Art, Literature, and Progress,** from the Creation of the World to the Present Time. The Continuation by W. D. HAMILTON, F.S.A. 3s.; cloth boards, 3s. 6d.

50. **Dates and Events in English History,** for the use of Candidates in Public and Private Examinations. By the Rev. E. RAND. 1s.

ENGLISH LANGUAGE AND MISCELLANEOUS.

11. **Grammar of the English Tongue,** Spoken and Written. With an Introduction to the Study of Comparative Philology. By HYDE CLARKE, D.C.L. Fourth Edition. 1s. 6d.

12. **Dictionary of the English Language,** as Spoken and Written. Containing above 100,000 Words. By HYDE CLARKE, D.C.L. 3s. 6d.; cloth boards, 4s. 6d.; complete with the GRAMMAR, cloth hds., 5s. 6d.

48. **Composition and Punctuation,** familiarly Explained for those who have neglected the Study of Grammar. By JUSTIN BRENAN 18th Edition. 1s. 6d.

49. **Derivative Spelling-Book:** Giving the Origin of Every Word from the Greek, Latin, Saxon, German, Teutonic, Dutch, French, Spanish, and other Languages; with their present Acceptation and Pronunciation. By J. ROWBOTHAM, F.R.A.S. Improved Edition. 1s. 6d.

51. **The Art of Extempore Speaking:** Hints for the Pulpit, the Senate, and the Bar. By M. BAUTAIN, Vicar-General and Professor at the Sorbonne. Translated from the French. 8th Edition, carefully corrected. 2s. 6d.

54. **Analytical Chemistry,** Qualitative and Quantitative, a Course of. To which is prefixed, a Brief Treatise upon Modern Chemical Nomenclature and Notation. By WM. W. PINK and GEORGE E. WEBSTER. 2s.

THE SCHOOL MANAGERS' SERIES OF READING BOOKS,

Edited by the Rev. A. R. GRANT, Rector of Hitcham, and Honorary Canon of Ely; formerly H.M. Inspector of Schools.
INTRODUCTORY PRIMER, 3d.

	s.	d.					s.	d.
FIRST STANDARD	0	6	FOURTH STANDARD	.	.	.	1	2
SECOND „	0	10	FIFTH „	.	.	.	1	6
THIRD	1	0	SIXTH „	.	.	.	1	6

LESSONS FROM THE BIBLE. Part I. Old Testament. 1s.

LESSONS FROM THE BIBLE. Part II. New Testament, to which is added THE GEOGRAPHY OF THE BIBLE, for very young Children. By Rev. C. THORNTON FORSTER. 1s. 2d. *⸎* Or the Two Parts in One Volume. 2s.

FRENCH.

24. **French Grammar.** With Complete and Concise Rules on the Genders of French Nouns. By G. L. STRAUSS, Ph.D. 1s. 6d.
25. **French-English Dictionary.** Comprising a large number of New Terms used in Engineering, Mining, &c. By ALFRED ELWES. 1s. 6d.
26. **English-French Dictionary.** By ALFRED ELWES. 2s.
25,26. **French Dictionary** (as above). Complete, in One Vol., 3s. ; cloth boards, 3s. 6d. *** Or with the GRAMMAR, cloth boards, 4s. 6d.
47. **French and English Phrase Book :** containing Introductory Lessons, with Translations, several Vocabularies of Words, a Collection of suitable Phrases, and Easy Familiar Dialogues. 1s. 6d.

GERMAN.

39. **German Grammar.** Adapted for English Students, from Heyse's Theoretical and Practical Grammar, by Dr. G. L. STRAUSS. 1s. 6d.
40. **German Reader :** A Series of Extracts, carefully culled from the most approved Authors of Germany ; with Notes, Philological and Explanatory. By G. L. STRAUSS, Ph.D. 1s.
41-43. **German Triglot Dictionary.** By N. E. S. A. HAMILTON. In Three Parts. Part I. German-French-English. Part II. English-German-French. Part III. French-German-English. 3s., or cloth boards, 4s.
41-43 **German Triglot Dictionary** (as above), together with German
& 39. Grammar (No. 39), in One Volume, cloth boards, 5s.

ITALIAN.

27. **Italian Grammar,** arranged in Twenty Lessons, with a Course of Exercises. By ALFRED ELWES. 1s. 6d.
28. **Italian Triglot Dictionary,** wherein the Genders of all the Italian and French Nouns are carefully noted down. By ALFRED ELWES. Vol. 1. Italian-English-French. 2s. 6d.
30. **Italian Triglot Dictionary.** By A. ELWES. Vol. 2. English-French-Italian. 2s. 6d.
32. **Italian Triglot Dictionary.** By ALFRED ELWES. Vol. 3. French-Italian-English. 2s. 6d.
28,30, **Italian Triglot Dictionary** (as above). In One Vol., 7s. 6d.
32. Cloth boards.

SPANISH AND PORTUGUESE.

34. **Spanish Grammar,** in a Simple and Practical Form. With a Course of Exercises. By ALFRED ELWES. 1s. 6d.
35. **Spanish-English and English-Spanish Dictionary.** Including a large number of Technical Terms used in Mining, Engineering, &c. with the proper Accents and the Gender of every Noun. By ALFRED ELWES 4s. ; cloth boards, 5s. *** Or with the GRAMMAR, cloth boards, 6s.
55. **Portuguese Grammar,** in a Simple and Practical Form. With a Course of Exercises. By ALFRED ELWES. 1s. 6d.
56. **Portuguese-English and English-Portuguese Dictionary.** Including a large number of Technical Terms used in Mining, Engineering, &c., with the proper Accents and the Gender of every Noun. By ALFRED ELWES. Second Edition, Revised, 5s. ; cloth boards, 6s. *** Or with the GRAMMAR, cloth boards, 7s.

HEBREW.

46*. **Hebrew Grammar.** By Dr. BRESSLAU. 1s. 6d.
44. **Hebrew and English Dictionary,** Biblical and Rabbinical ; containing the Hebrew and Chaldee Roots of the Old Testament Post-Rabbinical Writings. By Dr. BRESSLAU. 6s.
46. **English and Hebrew Dictionary.** By Dr. BRESSLAU. 3s.
44,46. **Hebrew Dictionary** (as above), in Two Vols., complete, with
46*. the GRAMMAR, cloth boards, 12s.

LONDON : CROSBY LOCKWOOD AND SON,

LATIN.

19. **Latin Grammar.** Containing the Inflections and Elementary Principles of Translation and Construction. By the Rev. THOMAS GOODWIN, M.A., Head Master of the Greenwich Proprietary School. 1s. 6d.

20. **Latin-English Dictionary.** By the Rev. THOMAS GOODWIN, M.A. 2s.

22. **English-Latin Dictionary;** together with an Appendix of French and Italian Words which have their origin from the Latin. By the Rev. THOMAS GOODWIN, M.A. 1s. 6d.

20,22. **Latin Dictionary** (as above). Complete in One Vol., 3s. 6d. cloth boards, 4s. 6d. *.* Or with the GRAMMAR, cloth boards, 5s. 6d.

LATIN CLASSICS. With Explanatory Notes in English.

1. **Latin Delectus.** Containing Extracts from Classical Authors, with Genealogical Vocabularies and Explanatory Notes, by H. YOUNG. 1s. 6d.

2. **Cæsaris Commentarii de Bello Gallico.** Notes, and a Geographical Register for the Use of Schools, by H. YOUNG. 2s.

3. **Cornelius Nepos.** With Notes. By H. YOUNG. 1s.

4. **Virgilii Maronis Bucolica et Georgica.** With Notes on the Bucolics by W. RUSHTON, M.A., and on the Georgics by H. YOUNG. 1s. 6d.

5. **Virgilii Maronis Æneis.** With Notes, Critical and Explanatory, by H. YOUNG. New Edition, revised and improved With copious Additional Notes by Rev. T. H. L. LEARY, D.C.L., formerly Scholar of Brasenose College, Oxford. 3s.

5* ———— Part 1. Books i.—vi., 1s. 6d.

5** ———— Part 2. Books vii.—xii., 2s.

6. **Horace;** Odes, Epode, and Carmen Sæculare. Notes by H. YOUNG. 1s. 6d.

7. **Horace;** Satires, Epistles, and Ars Poetica. Notes by W. BROWNRIGG SMITH, M.A., F.R.G.S. 1s. 6d.

8. **Sallustii Crispi Catalina et Bellum Jugurthinum.** Notes, Critical and Explanatory, by W. M. DONNE, B.A., Trin. Coll., Cam. 1s. 6d.

9. **Terentii Andria et Heautontimorumenos.** With Notes, Critical and Explanatory, by the Rev. JAMES DAVIES, M.A. 1s. 6d.

10. **Terentii Adelphi, Hecyra, Phormio.** Edited, with Notes, Critical and Explanatory, by the Rev. JAMES DAVIES, M.A. 2s.

11. **Terentii Eunuchus, Comœdia.** Notes, by Rev. J. DAVIES, M.A. 1s. 6d.

12. **Ciceronis Oratio pro Sexto Roscio Amerino.** Edited, with an Introduction, Analysis, and Notes, Explanatory and Critical, by the Rev JAMES DAVIES. M.A. 1s. 6d.

13. **Ciceronis Orationes in Catilinam, Verrem, et pro Archia.** With Introduction, Analysis, and Notes, Explanatory and Critical, by Rev. T. H. L. LEARY, D.C.L. formerly Scholar of Brasenose College, Oxford. 1s. 6d.

14. **Ciceronis Cato Major, Lælius, Brutus, sive de Senectute, de Amicitia, de Claris Oratoribus Dialogi.** With Notes by W. BROWNRIGG SMITH M.A., F.R.G.S. 2s.

16. **Livy : History of Rome.** Notes by H. YOUNG and W. B. SMITH, M.A. Part 1. Books i., ii., 1s. 6d.

16*. ———— Part 2. Books iii., iv., v., 1s. 6d.

17. ———— Part 3. Books xxi., xxii., 1s. 6d.

19. **Latin Verse Selections,** from Catullus, Tibullus, Propertius, and Ovid. Notes by W. B. DONNE, M.A., Trinity College, Cambridge. 2s.

20. **Latin Prose Selections,** from Varro, Columella, Vitruvius, Seneca, Quintilian, Florus, Velleius Paterculus, Valerius Maximus Suetonius, Apuleius, &c. Notes by W. B. DONNE, M.A. 2s.

21. **Juvenalis Satiræ.** With Prolegomena and Notes by T. H. S. ESCOTT, B.A., Lecturer on Logic at King's College, London. 2s.

GREEK.

14. Greek Grammar, in accordance with the Principles and Philological Researches of the most eminent Scholars of our own day. By HANS CLAUDE HAMILTON. 1s. 6d.

15,17. Greek Lexicon. Containing all the Words in General Use, with their Significations, Inflections, and Doubtful Quantities. By HENRY R. HAMILTON. Vol. 1. Greek-English, 2s. 6d.; Vol. 2. English-Greek, 2s. Or the Two Vols. in One, 4s. 6d.: cloth boards, 5s.

14,15. Greek Lexicon (as above). Complete, with the GRAMMAR, in
17. One Vol., cloth boards, 6s.

GREEK CLASSICS. With Explanatory Notes in English.

1. Greek Delectus. Containing Extracts from Classical Authors, with Genealogical Vocabularies and Explanatory Notes, by H. YOUNG. New Edition, with an improved and enlarged Supplementary Vocabulary, by JOHN HUTCHISON, M.A., of the High School, Glasgow. 1s. 6d.

2, 3. Xenophon's Anabasis; or, The Retreat of the Ten Thousand. Notes and a Geographical Register, by H. YOUNG. Part 1. Books i. to iii., 1s. Part 2. Books iv. to vii., 1s.

4. Lucian's Select Dialogues. The Text carefully revised, with Grammatical and Explanatory Notes, by H. YOUNG. 1s. 6d.

5-12. Homer, The Works of. According to the Text of BAEUMLEIN. With Notes, Critical and Explanatory, drawn from the best and latest Authorities, with Preliminary Observations and Appendices, by T. H. I LEARY, M.A., D.C.L.

THE ILIAD:	Part 1. Books i. to vi., 1s.6d.	Part 3. Books xiii. to xviii., 1s. 6d.
	Part 2. Books vii. to xii., 1s. 6d.	Part 4. Books xix. to xxiv., 1s. 6d.
THE ODYSSEY:	Part 1. Books i. to vi., 1s. 6d	Part 3. Books xiii. to xviii., 1s. 6d.
	Part 2. Books vii. to xii., 1s. 6d.	Part 4. Books xix. to xxiv., and Hymns, 2s.

13. Plato's Dialogues: The Apology of Socrates, the Crito, and the Phædo. From the Text of C. F. HERMANN. Edited with Notes. Critical and Explanatory, by the Rev. JAMES DAVIES, M.A. 2s.

14-17. Herodotus, The History of, chiefly after the Text of GAISFORD. With Preliminary Observations and Appendices, and Notes, Critical and Explanatory, by T. H. L. LEARY, M.A., D.C.L.
Part 1. Books i., ii. (The Clio and Euterpe), 2s.
Part 2. Books iii., iv. (The Thalia and Melpomene), 2s.
Part 3. Books v.-vii. (The Terpsichore, Erato, and Polymnia), 2s.
Part 4. Books viii., ix. (The Urania and Calliope) and Index, 1s. 6d.

18. Sophocles: Œdipus Tyrannus. Notes by H. YOUNG. 1s.

20. Sophocles: Antigone. From the Text of DINDORF. Notes, Critical and Explanatory, by the Rev. JOHN MILNER, B.A. 2s.

23. Euripides: Hecuba and Medea. Chiefly from the Text of DINDORF. With Notes, Critical and Explanatory, by W. BROWNRIGG SMITH, M.A., F.R.G.S. 1s. 6d.

26. Euripides: Alcestis. Chiefly from the Text of DINDORF. With Notes, Critical and Explanatory, by JOHN MILNER, B.A. 1s. 6d.

30. Æschylus: Prometheus Vinctus: The Prometheus Bound. From the Text of DINDORF. Edited, with English Notes, Critical and Explanatory, by the Rev. JAMES DAVIES, M.A. 1s.

32. Æschylus: Septem Contra Thebes: The Seven against Thebes. From the Text of DINDORF. Edited, with English Notes, Critical and Explanatory, by the Rev. JAMES DAVIES, M.A. 1s.

40. Aristophanes: Acharnians. Chiefly from the Text of C. H. WEISE. With Notes, by C. S. T. TOWNSHEND, M.A. 1s. 6d.

41. Thucydides: History of the Peloponnesian War. Notes by H. YOUNG. Book 1. 1s. 6d.

42. Xenophon's Panegyric on Agesilaus. Notes and Introduction by Lt. F. W. JEWITT. 1s. 6d.

43. Demosthenes. The Oration on the Crown and the Philippics. With English Notes. By Rev. T. H. L. LEARY, D.C.L., formerly Scholar of Brasenose College, Oxford. 1s. 6d.